COMMUNICATIONS EQUIPMENT OF THE GERMAN ARMY

1933-1945

CHARLES J. BARGER

COMMUNICATIONS EQUIPMENT OF THE GERMAN ARMY

1933-1945

PALADIN PRESS
BOULDER, COLORADO

Communications Equipment of the German Army: 1933-1945
by Charles J. Barger

ISBN 0-87364-534-0
Printed in the United States of America

Published by Paladin Press, a division of
Paladin Enterprises, Inc., P.O. Box 1307,
Boulder, Colorado 80306, USA.
(303) 443-7250

Direct inquiries and/or orders to the above address.

CONTENTS

PREFACE

The communications equipment employed by the Third Reich is a fascinating story. It is a subject of considerable interest to military historians, collectors and radio enthusiasts. Until this point there has been very little published on the topic. This volume, the first in a series, provides a thorough presentation of the signal communications equipment that was developed and used by the German Army between 1933 and 1945.

German radio equipment has been relatively rare right from the beginning of the capitulation of the Wehrmacht in 1945. Few examples were brought back to the Allied countries as war trophies because of practicality. Another, and perhaps the most significant reason is that the Allied occupation forces were ordered to demolish the equipment.[1] Not only did thousands of radios ended up under the sledge hammer but design documents and records also were destroyed or dispersed as well.

1. Ham Radio Magazine, August 1980

CREDITS AND ACKNOWLEDGEMENTS

Preparation of this volume first began during July 1985 when a large number of radio sets came my way as a result of purchasing German military equipment. These sets had only recently been found on a German military installation remaining in storage since the end of the war. It only goes to show that there are still items out there to discover even after forty years. I would personally like to thank the following individuals without whose help this book could not have been written: From West Germany I would like to thank Gunther Buchholz and Klaus Brase for assisting me in locating the equipment and sending it all the way to the United States. Shipping was a major obstacle due to the size and weight involved. In the United States a very special thanks goes to Art Obermeyer of St. Louis, Missouri. Art played a major role during the preparation of this book by sending documents and technical information. He personally worked on several items and completely restored the Torn. fu. d2 shown in this book. Although anonymous, the Army Signal personnel that wrote the technical bulletins during the war deserve special recognition.

The text of chapters 3-9 are reprinted in their entirety from the US Army Technical Bulletins. Wherever possible, I have tried to enhance these reports by adding in detailed photographs of the equipment.

The following individuals and organizations also rendered valuable assistance in the preparation of this book:

National Archives and Records Service
Zentralbiblothek der Bundeswehr
The Patton Museum of Cavalry and Armor
RZM Imports Collection
ChicTech Services - Electronic Typesetting, Graphics and Layout
Bob Griffen - Photography
Bill Frake
Chris Schneider

Photo Credits:

Arthur Obermeyer: 78, 79, 80

RZM Imports: 169, 170

Wilhelm Limpert Verlag: 2, 49, 50, 71, 72, 106, 172

U.S. Army: 39, 40, 54, 61, 90, 91, 92, 93, 98, 109, 112, 113, 128, 130, 142, 145, 147
155, 157

Bybdesarcgv: 103, 138, 157

Those photographs not listed shall be credited to the author and the organizations or individuals who prefer to remain anonymous.

GERMAN PRODUCTION CODES

This list, authenticated from the original "List of Manufacturers" Marks for weapons, ammunition and equipment, published by the Supreme Command of the Army Ordnance Department (Berlin 1944) delineates the Secret Codes for manufacturer's relative to signal communications equipment.

al — Deutsches Leucht - μ. Signalmittelwerk, Dr. Feistal K-G
Berlin - Charlottenburg 9 Kaiserdamm 44
Werk Schonhogen b. Trebbin Krs Teltow

bl — Radio H Mende & Co., Rundfunkapparatefabriken Dresden N 15, Industriegelände

bo — Eltz. Jng. Nikolaus, Rundfunkapparatefabrik
Wien 55 Gartengasse 14 16

bg — Brandt, Roland, Fabrik f. Radio Telefonie
Berlin So 36, Köpenickerstr. 62 - 7

bt — Radiowerk Horny A. - G Rundfunkapparatefabrik Wien 75 Südbahngasse

bμ — Henry & Co., Heinr., Unternehmen für Elektro - Akustik, Radio - μ. Kinotechnik,
Wien VIII/62 Schottenfeldgasse 39

bw — Minerva Radio, W. Wohlleber & Co.,
Rundfunkapparatefabrik, Wien 62, Zieglergasse

cd — Roederstein, Ernst, Rundfunkzubehör,
Berlin So 16, Wusterhausenstr. 16

cj — Neuberger, Josef, Fabrik electrischer Instrumente
München 25, Steinerstr. 16

cn — Borck & Goldschmidt, Mechanische Werkstätten
für Telegrafie und Telefonie
Berlin 0 17, Fruchtstr. 1 -2

ber — Reiner, Friedrich, Telefonfabrik
München, Jahnstr. 38.

bkc — Hauser & Co., Telefonfabrik,
München 42, Lutzstr. 25

bmm — Lumophon - Werke, Bruckner & Stark,
Fernsprecher; Radio - μ. Kühlschrank Fabr.
Nürnberg - 0., Schlosstr. 62 - 64

bou	Telefunken Gesellschaft für drahtlose Telegraphie m. b. 14 Werk Erfurt und Berlin - Zehlendorf.
bpt	Telefon - µ. Telegraphenfabrik - A.-G, Kapsch & Söhne Wien 87/XII, Johann Hoffmannplatz 9
bqa	Vereinigte Telefon µ. Telegrafen - Werke A.-G Wein 20, Dresdener Str. 75
bug	Süddeutsche Telefon - Apparate - Kabel - und Drahtwerke A. - G., Teka Dc Nürnberg, Nornenstr 33
bxf	Telefonbau & Normalzeit G.m.b.H., Berlin SW 68, Markgrafenstr. 76
bxo	Deutsche Telefonwerke und Kabel Industrie A. - G Berlin SO 36, Zeughofstr. 6 - 9
byj	Straßfurter Rundfunk - Gesellschaft m.b.H., Staßfurt/S2.
caw	Valvo Radioröhren G.m.b.H., Apparatefabrik Aachen, Aachen, Jagarstr.
cmy	Krüger, Rudolph, Telegraphen, Bau anstalt, Berlin SO 16, Michaelkirchstr. 41
cqp	Signalapparatefabrik Julius Kächer A. - G., Fabrik für Fernsprechapparate µ. Zubehör, Berlin SW 61, Belle - Alliancestr. 61
dej	Siemens - Elektrowärme G.m.b.H., Sörnewitz b/Meissen
dey	Kammere, A. & R., Telefon - R Apparaetebau G.m.b.H., München 9. Tassiloplatz 6
djx	Telephon - µ Telegraphenwerke Stöcker & Co., Leipzig W 31 Wehsmuthstr. 10
djy	Kötten & Cic. G.m.b.H., Fabrik für elektrotechn. Artikel, Berg. Gladbach
dmr	Lorenz A.- G., C., Telephone - µ Telegraphenwerke, Berlin - Tempelhof, Lorenzweg

dnb Borck & Goldschmidt, Mechanische Werkstätten f. Telefgraphie µ. Telefonie, Berlin 0 17 Frucht str. 1 -2

dnm Siedle & Söhne S. Telefon - µ Telegrafenwerke A. - G., Furtwangen/B

duk Telefunken Ges. f. drahtlose Telegraphie, Werk Neuhaus am Rennweg/Thür

dul Radioröhrenfabrik G.m.b.H, Rundfunkzubehörteile. Hamburg - Lockstedt 1. Maikowskialler

dyk Wieser Karl Feinmechanik Telefonbau München 5. Jekstattstr. 5

eaa Telefonbau und Normalzeit G.m.b.H., Frankfurt (Main) 17. Mainzerland Str. 134-142

emy Merk, Telefonbau A. - G. Freidrich München, Werngeuestr. 32

eob Mauz & Pfeiffer, Elektro - Apparatebau Stuttgart - Botnang, Alte Stuttgarterstr 45 - 47

epz Elektrotechnische Fabrik Weber & Co., K. - G., Kranichfeld/Thür.

egn Eichenouer, Fritz Fabrik Elektr. Spezialaktikel Kandel/Pfalz

erk Elektrotechnische Fabrik Eltefa G.m.b.H., Lechesten/Thür.

fsb Mikrofona Brüder Knotek, Telefon - µ. Radiotechnische Fabrik Werke Prag, Prag XIII. - Strasnice

fsc Mikrofona Brüder Knotek, Telefon - µ radiotechnische Fabrik Werk, Wal Meseritsch

fxk Telegrafia A.G., für Telegraphen - µ. Telephonerzeugung, Prog. 1. Narodni Trida 25

gec Fernsprech - µ Signalbaugesellschaft
Schüler & Verschoven,
Essen, Schäferstr. 7-9

gwf Becker, Otto, Fabrik elektrischer
Maschinen µ. Apparate,
Berlin - Schöneberg, Belzigerstr. 25

hve Carstens & Co., J
Batterie - µ. Elementefabrik,
Hamburg 20, Falkenried 74

hwe Hecht, Fritz Maschinnetelegrafenfabrik
Keil, Lübecker Chaussce 27a

hwh Hönnecke, Martin, Fabrik elektr. Specialmaschinen
Leipzig C 1 Leplastr. 10a f.

jhk Strumpf, Herman
Berliner Telefonfabrik
Berlin SO 36, Adalbertstr. 92

jmq Telefon - µ. Elektrizitätsges - G.m.b.H.,
Brühn 17, Körnenrgrasse 12

jkn Allgemeine Telefon - Fabrik G.m.b.H.,
Hamburg 36, Fuhlentweiete 51-53

lut Teleradio A. - G., für Telefonie - µ. Telegraphie,
Bern, Bundesgasse 10.

mmz C. Lorenz A.G., Berlin - Tempelhof.

npv Fabrik elektrischer Steuerapparate
Rudolf Knote, Leipzig W 34
Dieskauster, 131

nqd Fischer & Freund
Berlin - Pankow, Brechmestr. 21 - 22

The use of codes when applied to communications equipment differs somewhat than other materials produced during the war. For example, of all the radio sets illustrated in this volume not one of the receivers or transmitters is marked with a three letter code. A surprising number of commercial trademarks, logos and corporate symbols were often used by German firms in place of the assigned three letter codes. I am unable to explain why the radio sets lack the coding. For the sake of continuity we know that the main contractors for radio sets were Lorenz, Telefunken and Siemens.

I

The Feldfernsprecher 33

THE FELDFERNSPRECHER 33

1. DESCRIPTION: The **Feldernsprecher** 33 or field telephone 33 is a general purpose field telephone designed for wire communications. It can also be used to remote off the Torn. fu. d2 and Torn. fu. b1, F series radio sets.

2. PERFORMANCE CHARACTERISTICS:

a. Dimensions and data:

Length: .... 28 cm
Width: .... 10 cm
Height: .... 21 cm

Weight:
a. upper and lower case only .... 1.65 kg
b. carrying strap .... .21 kg
c. battery .... .50 kg
d. handset .... .53 kg
e. switchboard cable .... .07 kg
f. apparatus parts with induction magnetic handcrank .... 2.88 kg

Total weight: .... 5.84 kg

b. The following components make up the set:

1 - handset
1 - switchboard cable
1 - induction crank
1 - shoulder strap
1 - 1.5 volt battery

Extra accessories available but not issued with set:
1 - headphone
1 - **Kehlkopf** or larynx microphone

3. GENERAL: The set is housed in a Bakelite plastic case approximately 3/16 inch thick. The case, consisting of 2 parts upper and lower, are hinged together and fasten with a self locking mechanism (figure 1). On the top of the upper case are 2 white panels. The panel on the left is used to record information while the panel on the right is inscribed with the German Phonetic alphabet (figure 2). An oval shaped dust cover can be swung open to give access to 2 input jacks. The jacks are needed when using another field telephone or central switchboard. The oval shaped grill on the lower front of the case is designed to enhance audibility of the bell. On the lower right side of the case is another access plate which protects the induction magneto (figure 3). Sockets are located on each side of the lower case along the edge for installing the carrying strap.

Pressing in on the latch opens the top cover. On the inside of the lid there are two circuit schematics and a wide leaf spring which keeps the hand set and keeps it from moving around (figure 4). Along the edge of the left side of the lower case is a rubber strip to keep the wires from being pinched when the case is shut. On the top left of the instrument panel is a compartment from the storing the switchboard cable. On the lower left of the instrument panel is the battery box. On the right side of the instrument panel are the white proof test button, marked **Pruft Taste**, the input jacks for headphones, and a recess for the storage of the induction crank, (figure 5). The handset is stored on top of the instrument panel with the mouthpiece

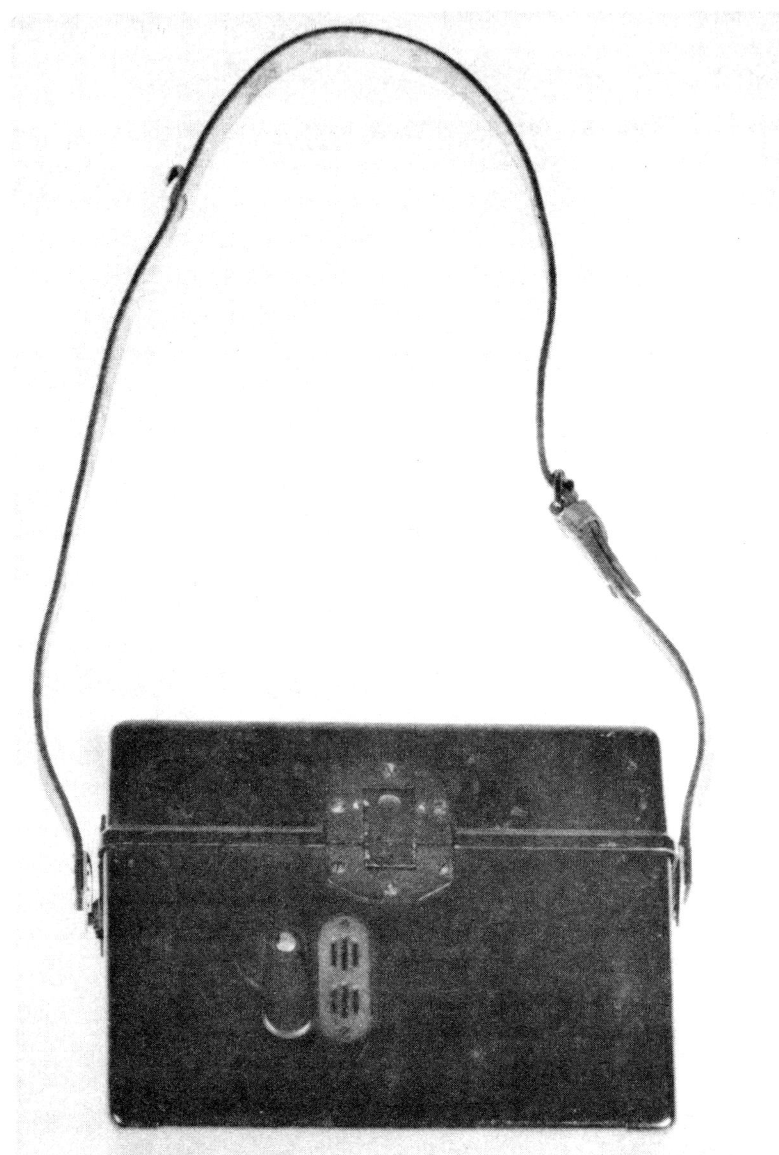

Figure 1. The Feldfernsprecher 33 with carrying strap attached. The small hook on the carrying strap is for hanging the handset.

Figure 2. All field telephones have the writing block and phonetic alphabet plates. The writing block is essential for writing down information such as codes, distant stations, etc. The phonetic alphabet is used in standard radio telephone procedures when communicating.

Figure 3. The handset can be positioned in a variety of ways. Note that the dust cover in the lower left of the photo has been opened for access to the input jacks for the connecting cable.

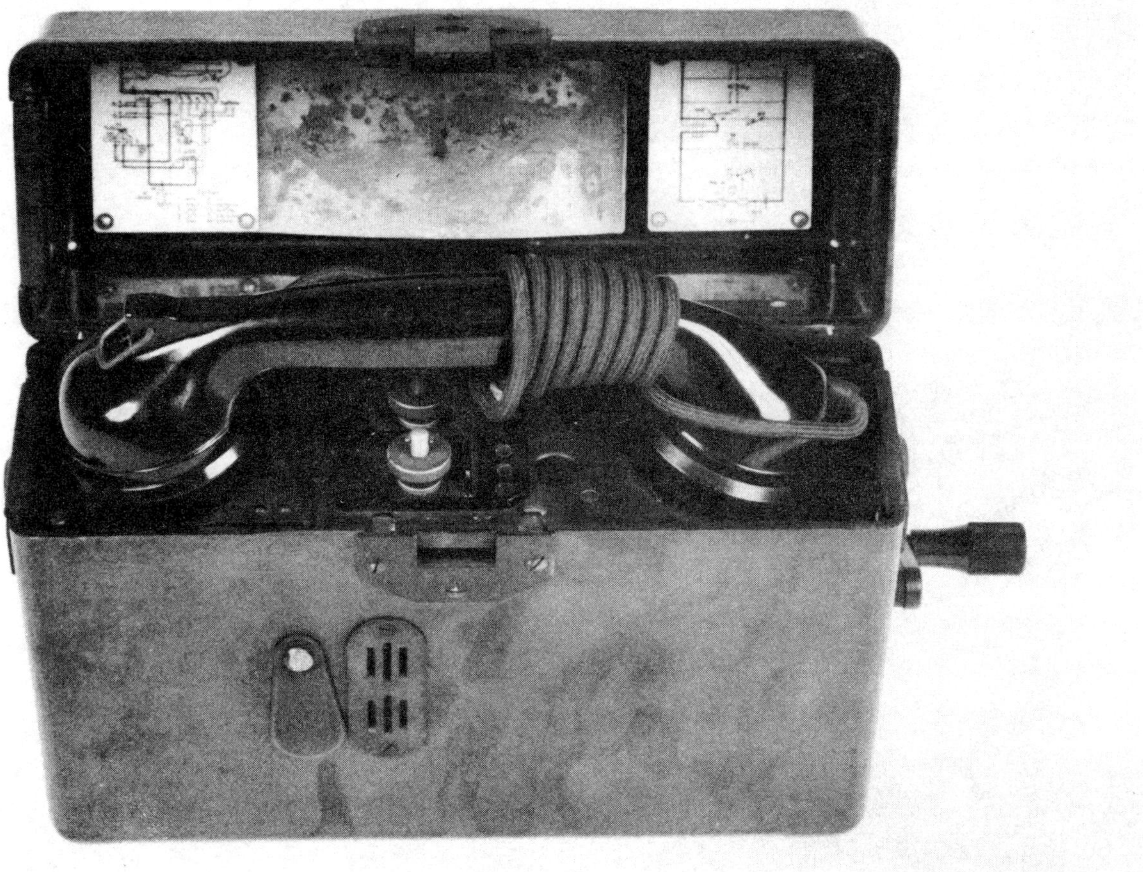

Figure 4. Front view showing the handset in the storage position. The induction crank has been installed as shown at the right side of the photograph.

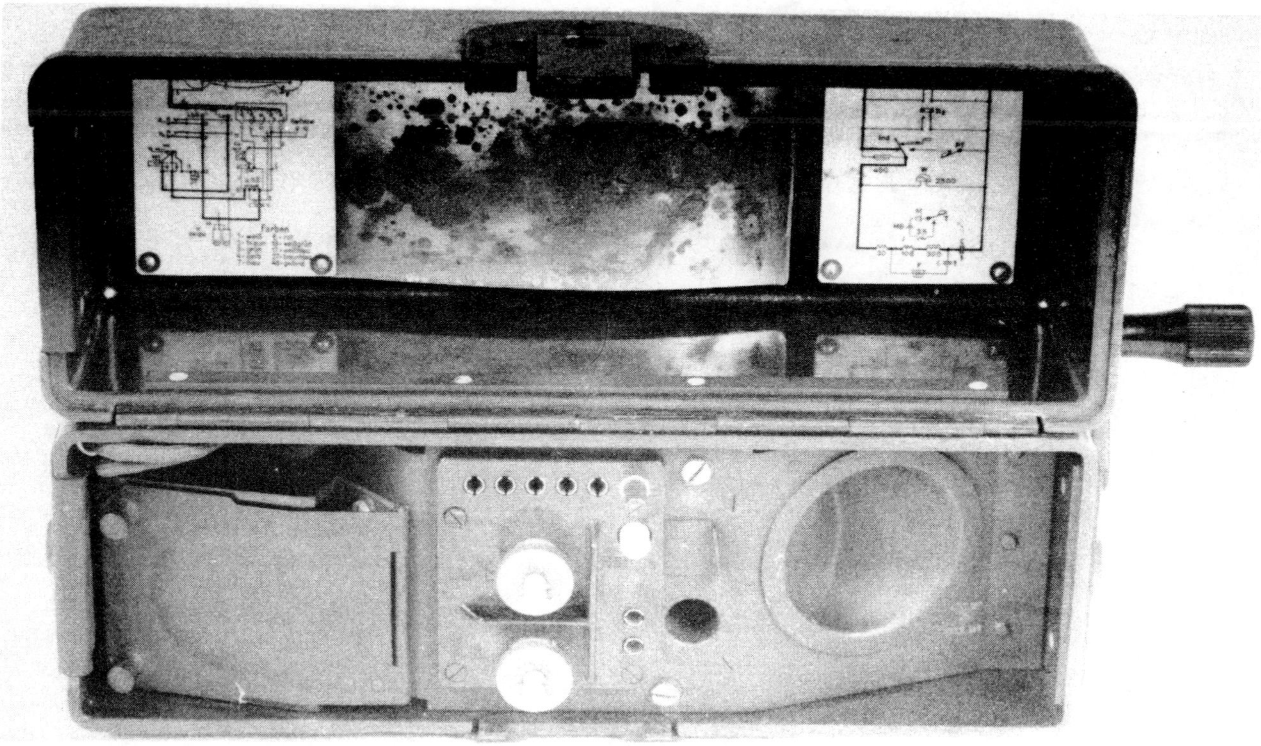

Figure 5. Top view of the instrument case with the handset removed. In the lower section from left to right is the battery box, terminal binding post, white proof test button, input jacks, and storage recess for the induction crank and handset. The handset will only fit into the case one way. The upper section of the illustration shows the schematics and pressure spring which holds the handset in place when the lid is closed.

Figure 6. The handset cord and connection block. The mouthpiece is shaped so that your voice will be directed into the microphone and not carry. Note the five pin connector block with schematic.

resting in the recess adjacent to the proof test button. To the right of the compartment for the switchboard cable is a five pin input jack for the handset plug. The five prong plug has a ring on the top to ease removal when wearing gloves. Inscribed on the side of the plug is a schematic showing how the prongs are wired. No: 1 and 5 are for the transmitter and talk switch, 2 and 4 are for the receiver (figures 6 & 7).

Figure 7. This illustration shows the handset with the internal components removed.

4. PLACING THE FIELD TELEPHONE INTO OPERATION.

a. Loosen the screw terminals on the top center of the instrument panel marked L (line a) and Lb/E (line b Edre or ground). Visually check the threaded post when loosening the screws from a small hole approximately 3/16 inch in diameter which runs thru the width of the screw. Run a length of communication wire (commo wire) to each terminal and insert it thru the hole. Take a turn around the post and tighten the screws down on the wire. Ensure that the wire has enough of the plastic coating stripped off the end to get a good electrical connection. Run the free end of the wire to the other phone and install the same way.

b. Open the lid on the battery box and insert a dry cell 1.5 volt battery into the battery holder. Wire the battery wire leads to the battery terminal post and close the battery box cover.

c. Remove the inductor hand crank from the holder, insert it thru the access opening on the lower right side of the bottom case and screw it into the induction magneto.

d. Remove the handset and ensure the 5 prong plug is connected to the input jack on the top of the instrument panel. Note: it will only go in one way.

e. Grasp the inductor hand crank and turn in vigorously in a clockwise fashion. This will ring the bell at the distant station. When you hear the other party, press the talkswitch bar on the handset and speak into the mouthpiece.

f. A white proof test (**Pruft Taste**) button located on the right side of the instrument panel is provided to test the line for breaks in the wire. To test the line, press down on the button while turning the induction crank. If your bell doesn't ring the line has a break in the wire and a runner will have to be sent out to inspect or repair the line.

g. It is possible to operate 2 phones together at one location by using the **Vermitlungsschnur** switchboard cable (figure 8) which is issued with each set. When a third party wishes to listen in on a call, place an additional field telephone near an operating set, remove the switchboard cable from the storage compartment, slide the access cover panel open on the front of the lower case and plug the cable into the input jack. When inserting the cable, be sure and connect the cables from bottom to bottom input jacks when using 2 phones. If more than 2 phones are used at one location run the cables from bottom to top or in series (see figure 9). The switchboard cable is also necessary when using the field telephone with the central switchboard.

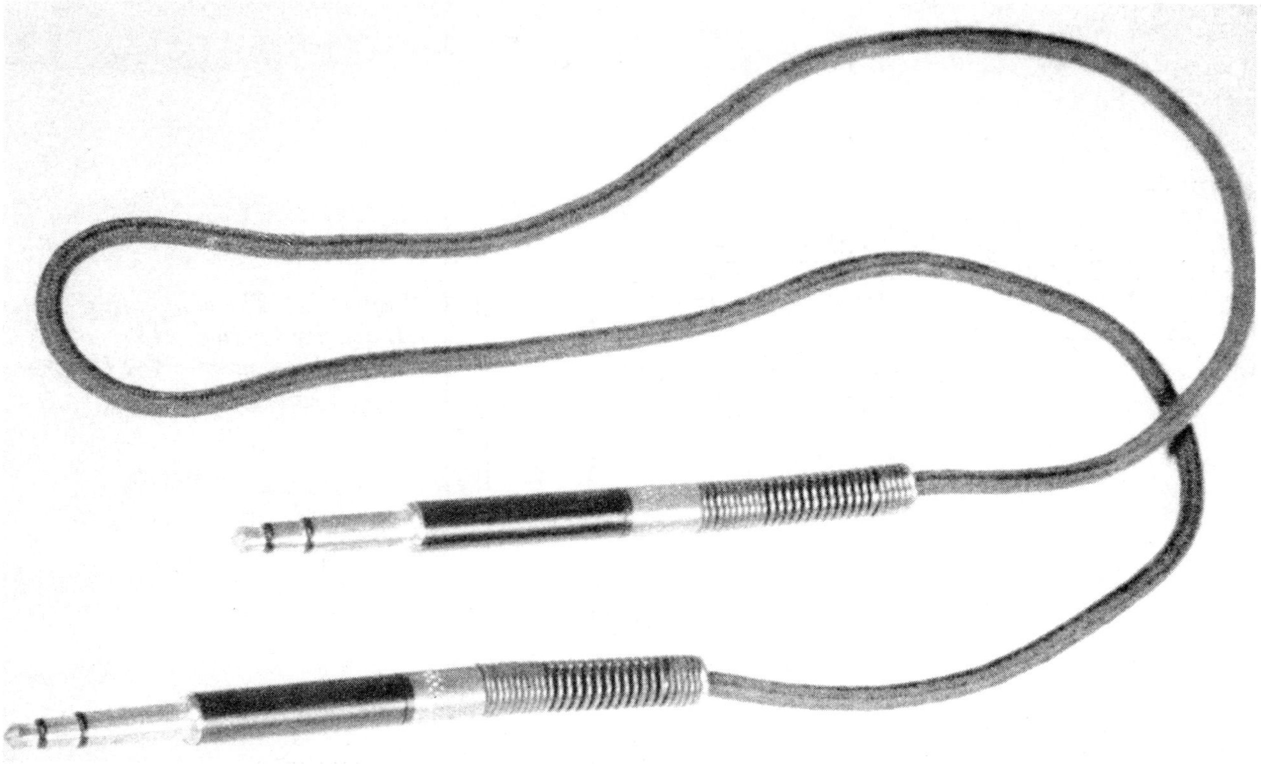

Figure 8. The connection cable is issued with each set. Cord lengths vary. This particular cord is 18 inches long.

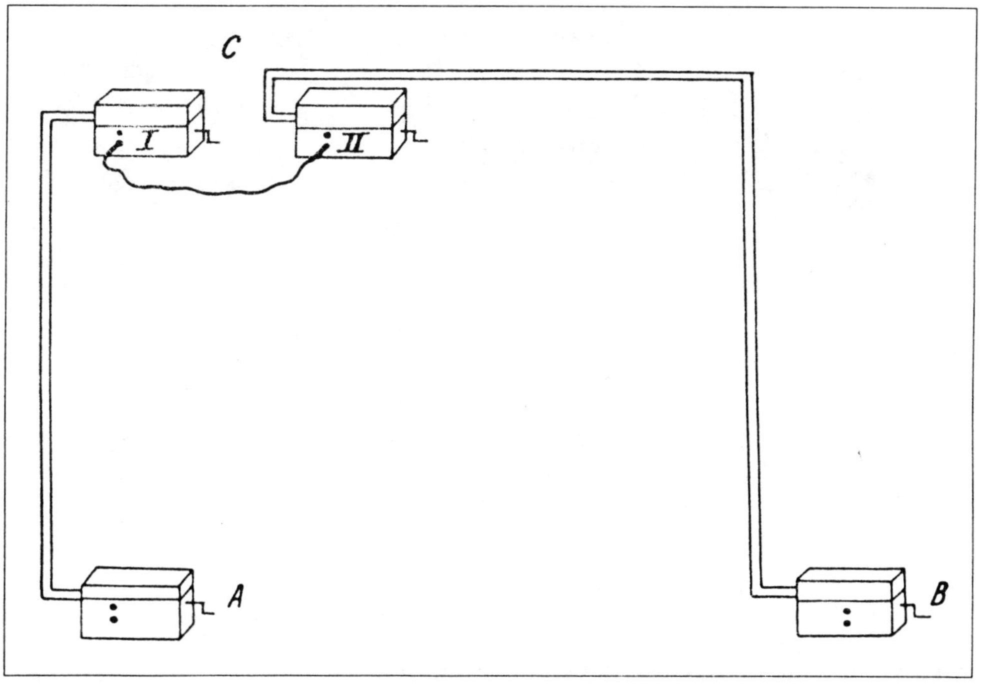

Figure 9a. The connection cable is being used in this illustration to link two phones together. The complete network now having four telephones on one line.

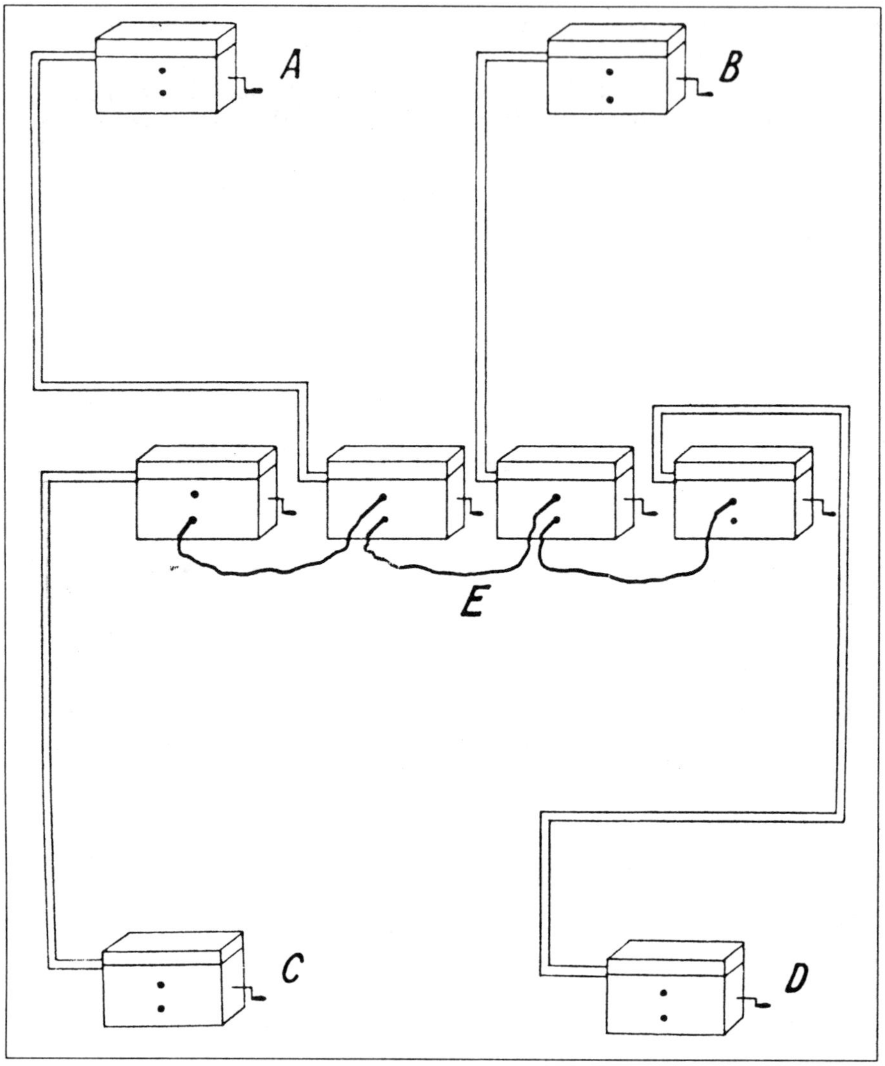

Figure 9b. The connecting cables at E are used between four telephones creating a central. Any telephone at E can now speak to A - D.

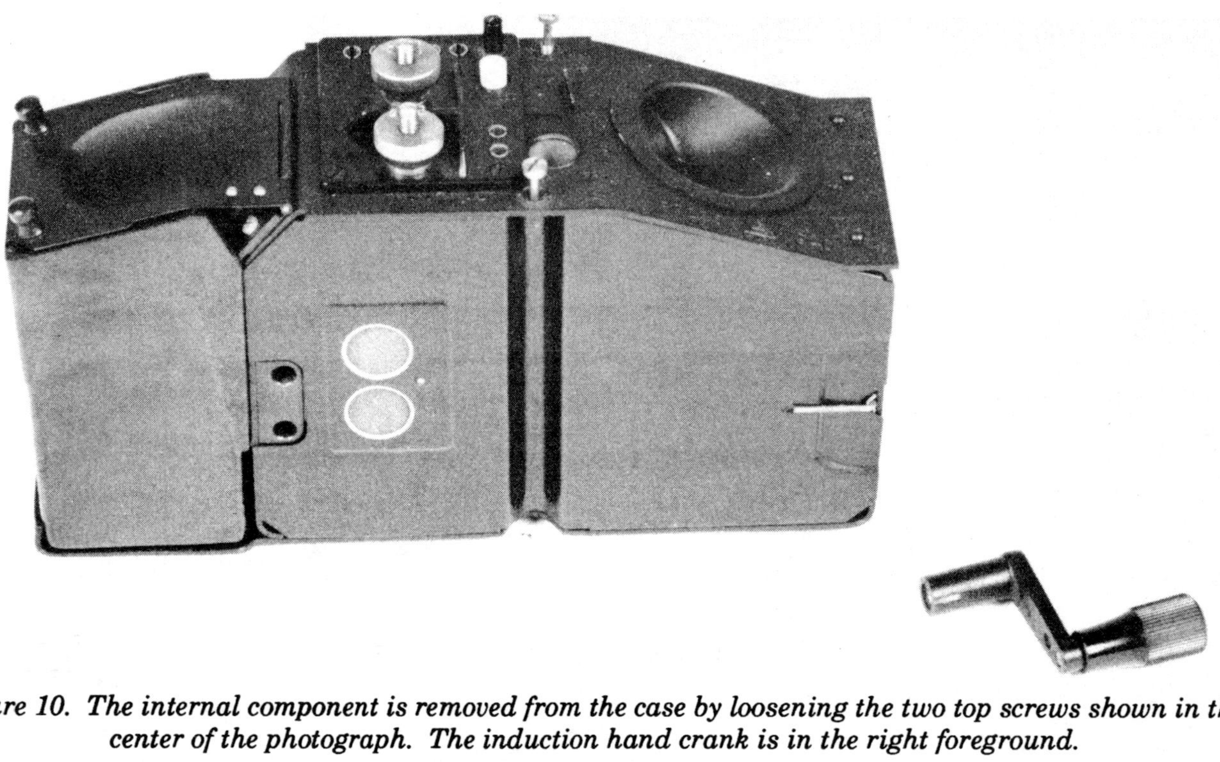

Figure 10. The internal component is removed from the case by loosening the two top screws shown in the center of the photograph. The induction hand crank is in the right foreground.

Figure 11. This view shows the internal components with side panel removed. Visible from left to right are the battery box, bell, and induction magneto. The side panels are locked in place by a wire bar, which is pushed up or down to release tension.

Figure 12. This illustration shows the opposite side.

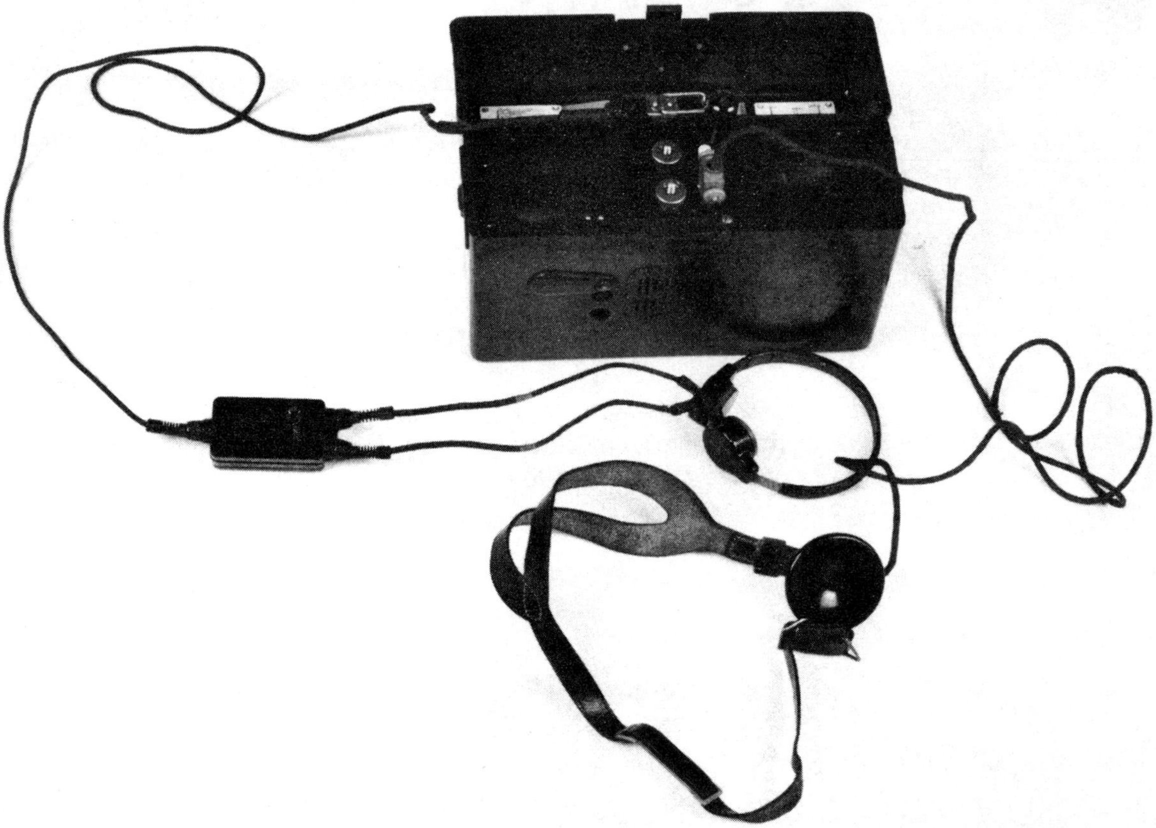

Figure 13. The illustration shows a kehlkopf or throat microphone installed with a headset. This headset is designed specifically for use when wearing a steel helmet. It can also be used with radio equipment depending on the type of plug end.

II

The Switchboards

THE SWITCHBOARDS

1. GENERAL: The field switchboard is a small portable unit capable of handling 10 incoming telephone lines. The set is housed in a wooden case with provisions for installing a carrying strap into the sockets located on each side (figure 1). A sheet metal cover is located on the top of the unit to protect the terminal post. The bottom of the set opens up for storing the cables. The **feldfernsprecher 33** must be used with the set in order to provide a means of ringing stations and talking.

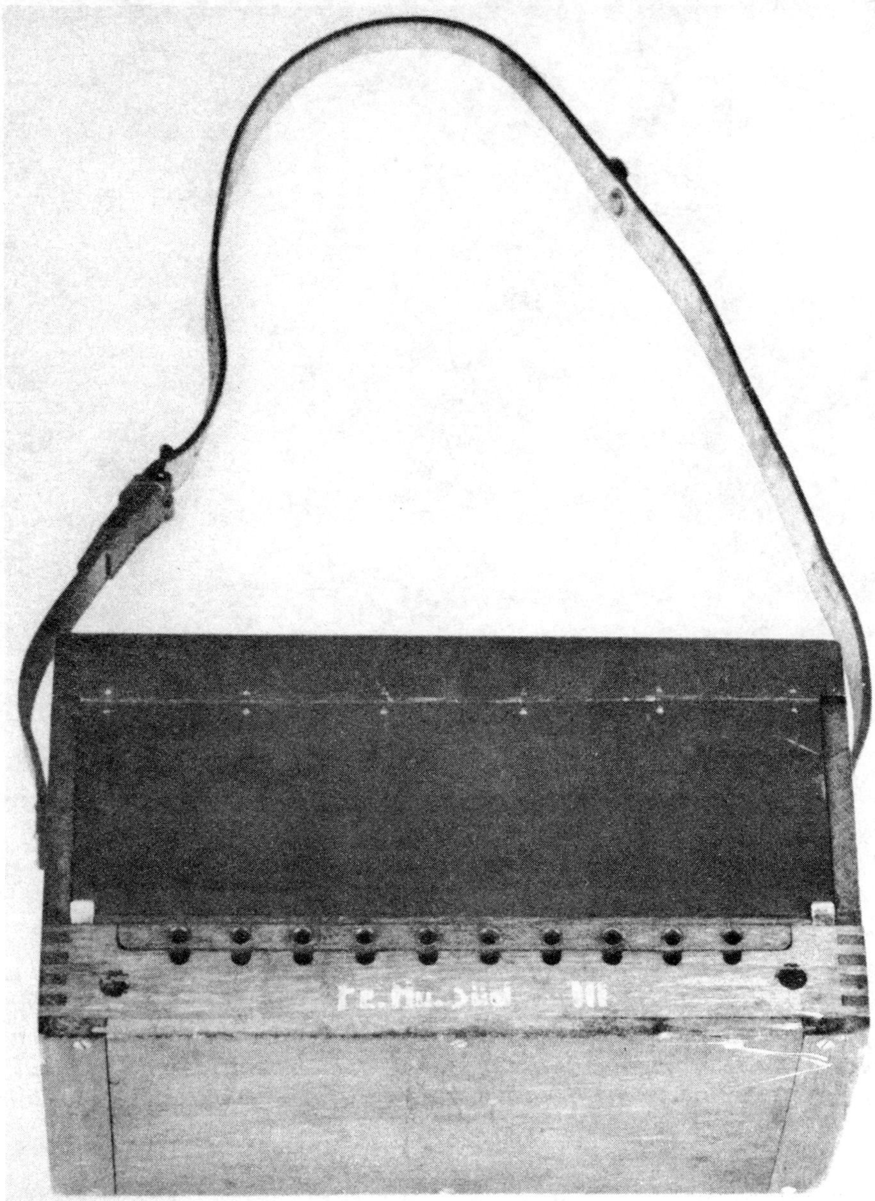

Figure 1. The 10 line switchboard in carrying configuration The carrying strap fastens into the sides of the case in the same manner as the field telephone.

2. PERFORMANCE CHARACTERISTICS:

Length: .... 14"
Width: .... 6"
Height: .... 7 3/4"
Weight Overall: .... 25 lbs.

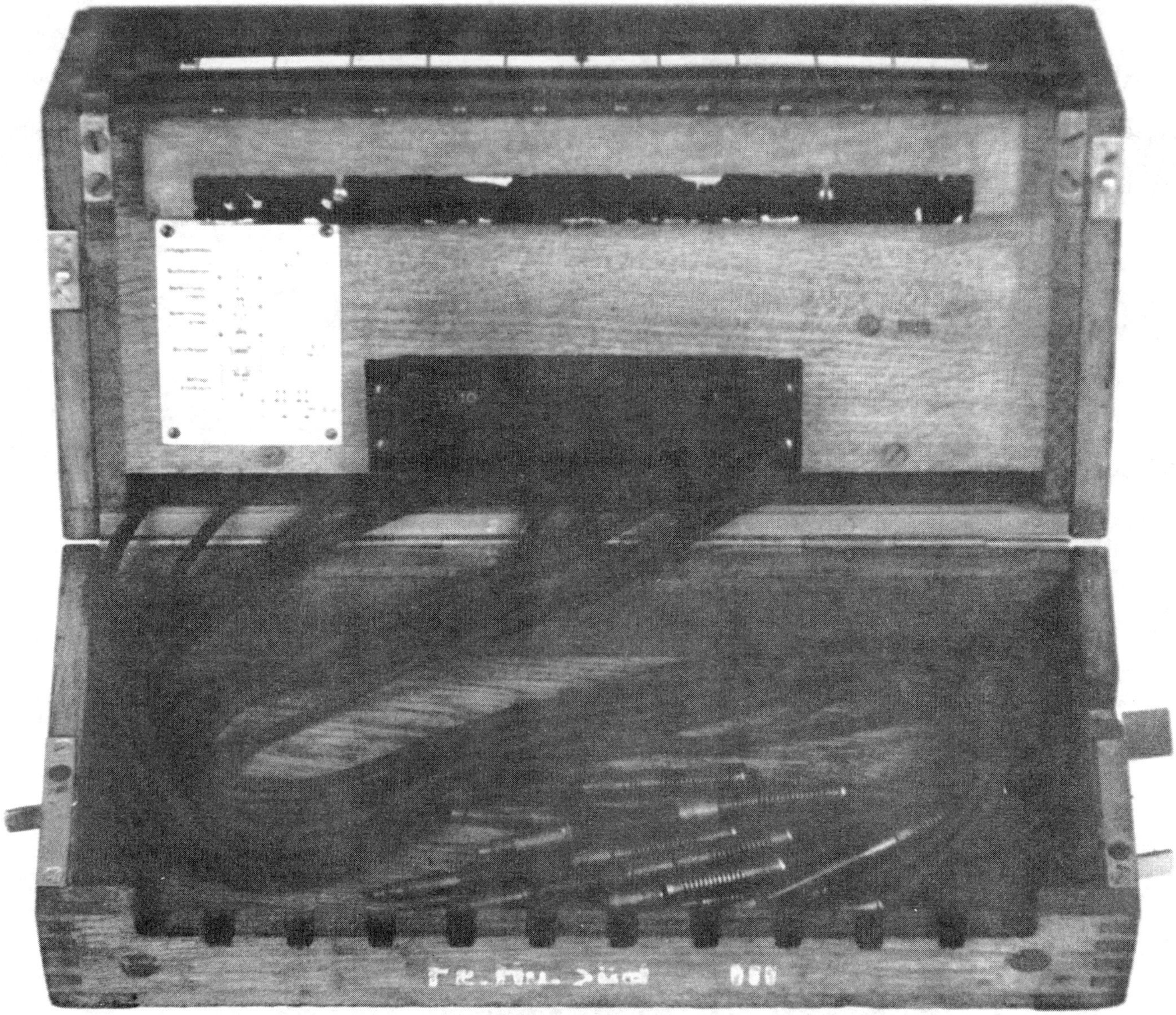

Figure 2. The cables are stored in the bottom of the case. Note the connection rack and wiring schematic. The connection rack (black rectangular box) is used as an alternate means of wiring the incoming lines. A 30 foot cable with a 30 pin plug end fits into the box. The free end is then run out away from the case thru the small access door shown at the lower right side of the photograph. The field phones would then be wired into the cable. This alternate method of wiring the set would be useful when set up in a building or entrenched position.

3. PLACING THE SET INTO OPERATION:

a. Remove the sheet metal top cover by loosening the screw on the center of the panel (figure 1).

b. Open the bottom compartment and remove the free ends of the connection cables (figure 2). Holes are located on the front of the instrument panel to hold the cables below their corresponding input jacks (figure 3). Loosen the metal bar over the front of the call flaps so they can drop.

c. Position a field telephone beside the switchboard and run a pair of wires from the field telephone terminals to the terminals on the switchboard marked **Abfr**. Place the field telephone into operation. The switchboard cable must be plugged into the front of the field telephone, this is to allow the operator to answer incoming calls.

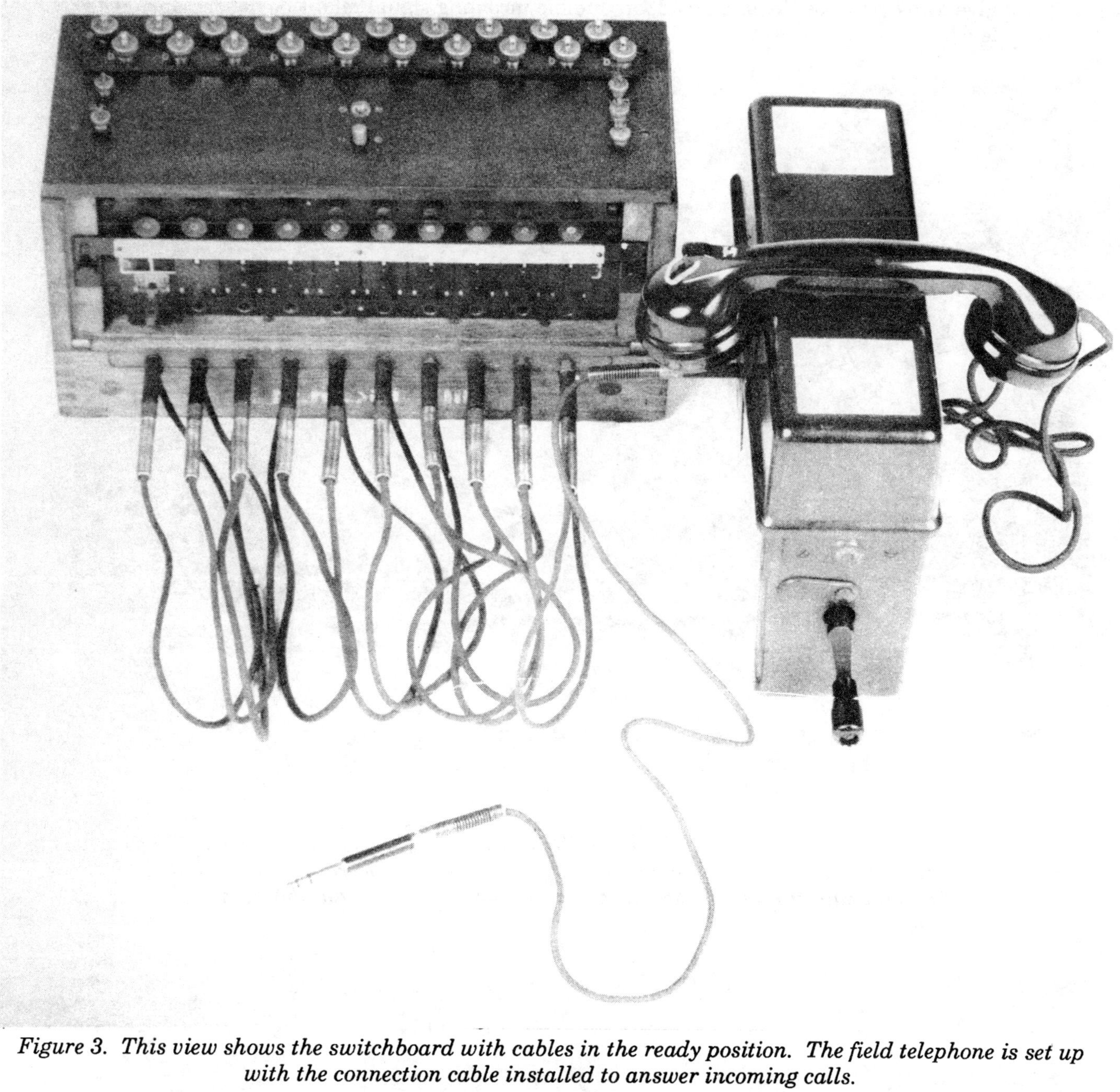

Figure 3. This view shows the switchboard with cables in the ready position. The field telephone is set up with the connection cable installed to answer incoming calls.

d. Attach incoming lines to the terminals numbered 1-10 using a and b rows for each telephone. The set can accommodate up to 10 different stations. The white plastic strip fastened to the bar below the call flaps can be used to pencil in names etc. The set is now ready for use (figures 3 & 4).

e. When the distant stations call in, the ringing current causes the **anrufklappe** (call flap) to drop. Take the loose end of the switchboard cable at the field telephone and answer the call by plugging it into the input jack. If the caller desires to speak to one of the other stations, take the callers cable, put it into the input jack of the desired station and push in the black button above the jacks. Ring the desired party using the field telephone. After ringing, press the red button to release the ringing connection key.

f. The only incoming signal the operator has is when the ringing current drops the call flap. If an audible signal is desired, a battery in series with a buzzer is hooked to terminals WB and W, W meaning Wecher and WB Wecher Batterie. The absence of, but provision for, the audible signal are twofold. First, in certain tactical situations, noise discipline might be important and incoming calls could give away your positions. Second, an audible incoming signal might be needed during high noise levels.

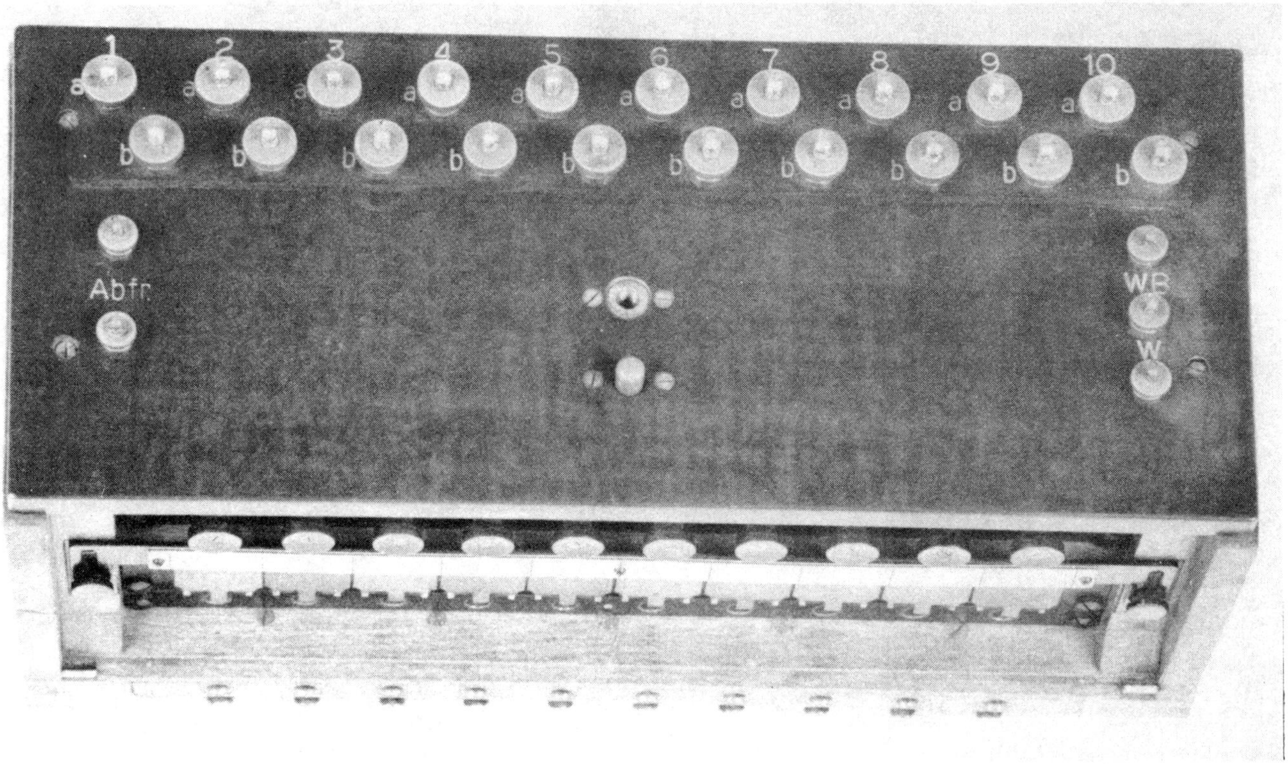

Figure 4. Top view of the switchboard. The numbered/lettered rows of terminal binding post show where to wire in the telephones. The 2 terminals at left marked Abfr is where the operator wires his field telephone. The terminals at right marked W-WB are for hooking up an audible signal.

Figure 4a. The instrument with cover removed. The white incremented bar across the front is raise or lowering by thumb screws. Prior to operation this bar must be placed in the upward position to allow the call flaps to drop. The operator can write in information on the panels to quickly identify stations. When the call flaps drop, a number will be visible which corresponds to the lines wired numerically on the top of the case.

Figure 5. Rear view of the case. Note the data plate at lower center.

Figure 6. A different type of switchboard. This set has its own provisions for ringing and answering calls without the use of a field telephone. This type of set has no way of keeping the cables from being entangled unlike the other model.

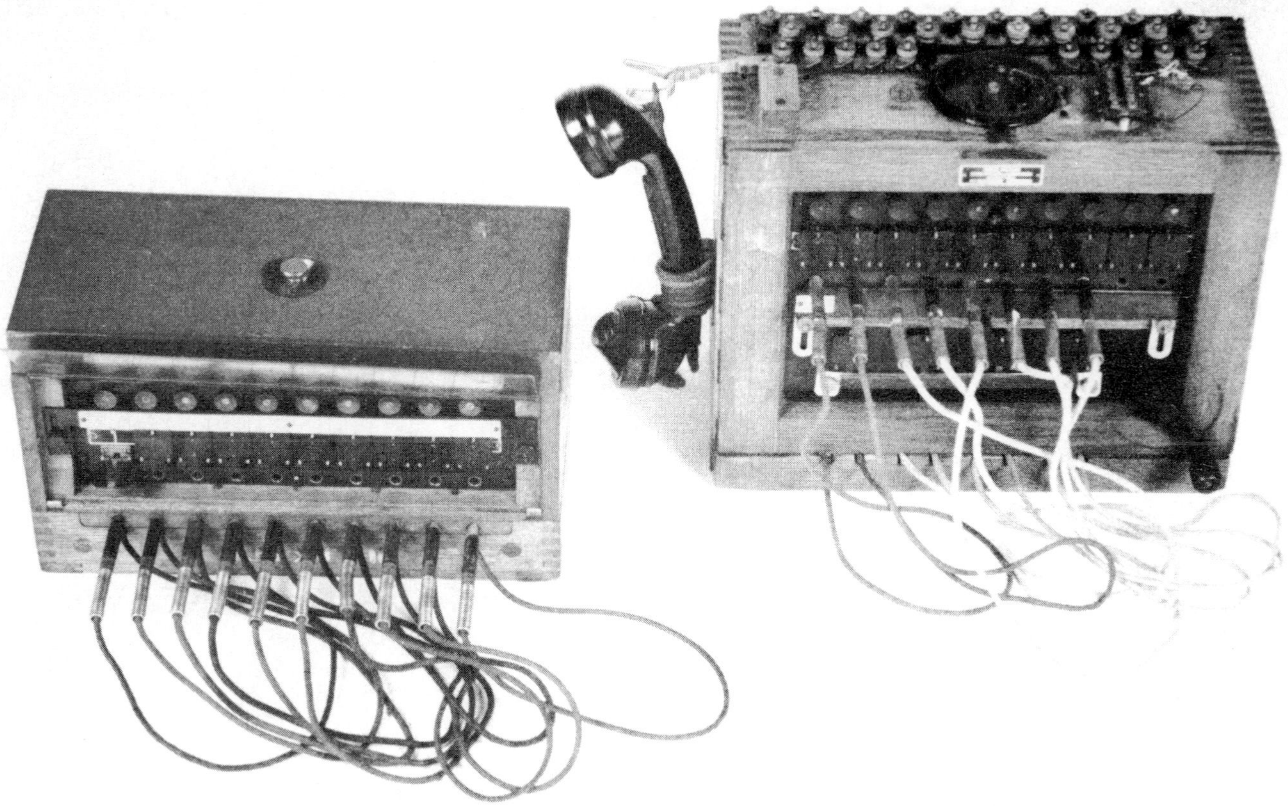

Figure 7. A comparison photo of two different models. The set at the left is more lightweight and compact, making it ideal for forward units. The larger set is more practical at higher echelons, and in fixed installations.

III

The Feld fu. b and c

THE FELD FU. b and FELD. FU. c

1. DESCRIPTION: The set is housed with all accessories in a molded plastic case, which is ribbed for additional strength. It can be carried and operated by one man so that both his hands are free. The accessories, as shown at figure 1, supplied with the set are:

1 – Two-section rod antenna, German type b for Feldfu. b and type c for Feldfu. c.
1 – 2.4-volt nickel-iron-storage battery, German type 2.4 NC 28.
1 – Remote control unit, German type b
1 – Remote Control cord, German type b
2 – Headsets, German type f
1 – Throat microphone, German type c

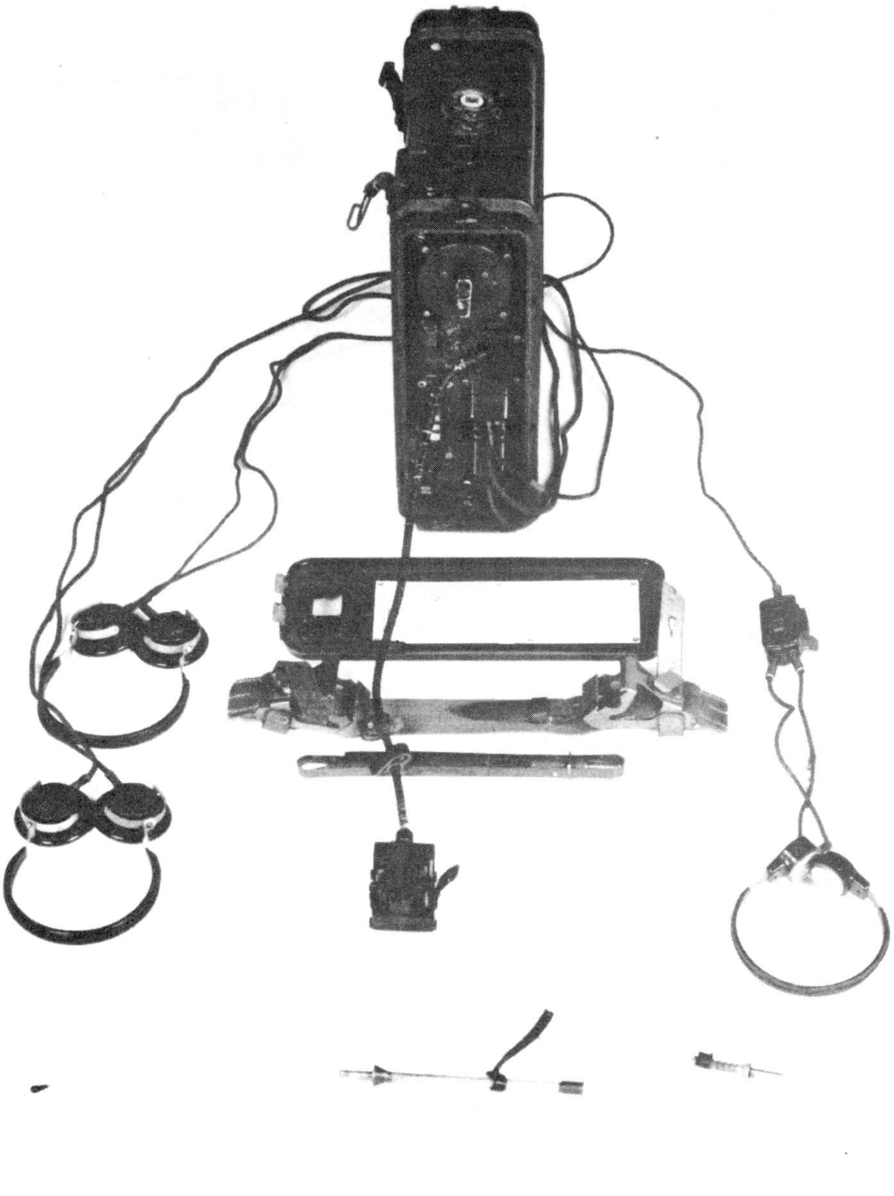

Figure 1. Feld fu. b. with accessories.

2. PERFORMANCE DATA: The following table lists the performance data of the sheet:

PERFORMANCE DATA

Frequency Range:

Radio Set	Transmitting	Receiving
Feldfu. b.	100-119 mc	100-119 mc
Feldfu. c.	120-138 mc	120-138 mc

Figure 2. Close up of the instrument panel. From top to bottom: frequency dial, remote control unit, battery life indicator, input jacks, data plate and on-off switch.

3. CONTROLS: The more important controls on the front panel and a brief statement of their functions are listed below (figures 2 - 4).

a. The **Frequenz-Einstellung** (tuning control) is a frequency selector divided into 30 channels with indents for indexing the pointer at each channel.

b. The **Frequ.-Prufg**. (calibration trimmer) is a trimmer capacitor used to align the receiver for net operation.

c. The **Lautstarke** (volume) is the receiver volume control.

d. The **Empf.-Nachstimmung** (receiver fine tuning) is a vernier tuning control for peaking the receiver on each channel.

e. The **Spannung** normal—**Sammler bereitst** (voltage normal—battery low) is a neon lamp that indicates the battery condition when the small check button is pressed.

f. The **Ein Aus** (on off) is a power switch used to turn the set on and off. This switch can be operated with the front cover on.

g. The microphone switch is a send-receive switch in the microphone cord which operates two relays in the set to change from transmit to receive and vice versa. This switch is marked **E.** (receive) and **S.** (transmit), and can be fastened by means of a spring clip to the front of the operator's uniform.

4. PREPARATION FOR USE:

a. Portable Operation.

(1) Make sure the **Ein Aus** (on off) switch is in the Aus (off) position.

Figure 3. This photo shows close up detail of the on-off switch, data plate, battery-voltage indicator button, and input jacks for one microphone and two headphones.

(2) Open the back cover of the set and remove the accessories from the case (figures 5 & 6).

(3) Make sure the battery terminals and the lugs of the battery leads are clean. Connect the positive battery lead to the red post of the battery, and the negative lead to the blue post of the battery. Strap the battery in place and close the back cover.

(4) Assemble the two-section antenna and plug it into the receptacle on top of the case. Make sure the top section is straight.

(5) Remove the front cover, slide the remote control unit to the left, and remove it from the panel. Plug the remote control cord into the remote control socket on the panel, and attach the other end to the remote control unit. This unit may now be fastened to the operator's belt (figures 7 - 8).

(6) Plug the throat microphone into the jack marked **Mikrofon** (microphone) and a headset into one of the jacks marked **Fernhorer** (headphones). Any high-impedance headset or single-button carbon microphone may be used if fitted with the proper plug.

(7) Place the microphone, headset, and remote control cords through the slot provided for them (figure 2, the slot is adjacent to the voltage–battery indicator).

(8) Set the **Ein Aus** (on off) switch to **Ein** (on).

(9) Press the small check button just below the neon indicator windows labelled **Spannung** normal (voltage normal) and **Sammler bereitst** (battery low). If both the indicators glow, the battery is well charged. If only the lower indicator glows, the battery is weak and will only operate the set for approximately 1/2 hour before recharging is necessary. If neither indicator glows, either the battery is discharged or the vibrator is defective.

Figure 4. Close up view of the frequency dial and remote control unit. The small access hole near the upper right base screw is for adjusting the calibration.

b. Stationary Operation. Preparation of the set for stationary operation is the same as for portable operation, except that the remote control unit need not be removed.

5. OPERATION:

a. Portable. After the set has been prepared for operation as directed in paragraph 4, proceed as follows (refer to figures 2 - 4):

(1) Set the **Frequenz-Einstellung** (tuning control) to the desired frequency channel.

(2) Set the microphone switch to **E.** (receive) position.

(3) Replace the front cover, and set the **Ein Aus** (on off) switch to **Ein** (on) by lifting the hinged tip of the knob and turning it to the right.

(4) Turn the **Lautstarke** (volume) control to the right until a rushing sound is heard in the headset, or until the desired signal is heard.

(5) Adjust the control marked **Empf.-Nachstimmung** (receiver fine tuning) for maximum volume.

(6) To transmit, set the microphone switch to the **S.** (transmit) position and start speaking.

b. Stationary. Stationary operation is the same as portable operation, except that the front cover may be left off.

6. CALIBRATION.

a. If the station cannot be tuned in by means of the **Empf.-Nachstimmung** (receiver fine tuning) control, it will be necessary to readjust the **Frequ.-Prufg.** (calibration trimmer).

Figure 5. Rear view of the case. The battery is installed in the compartment at the upper right, while accessories are carried below.

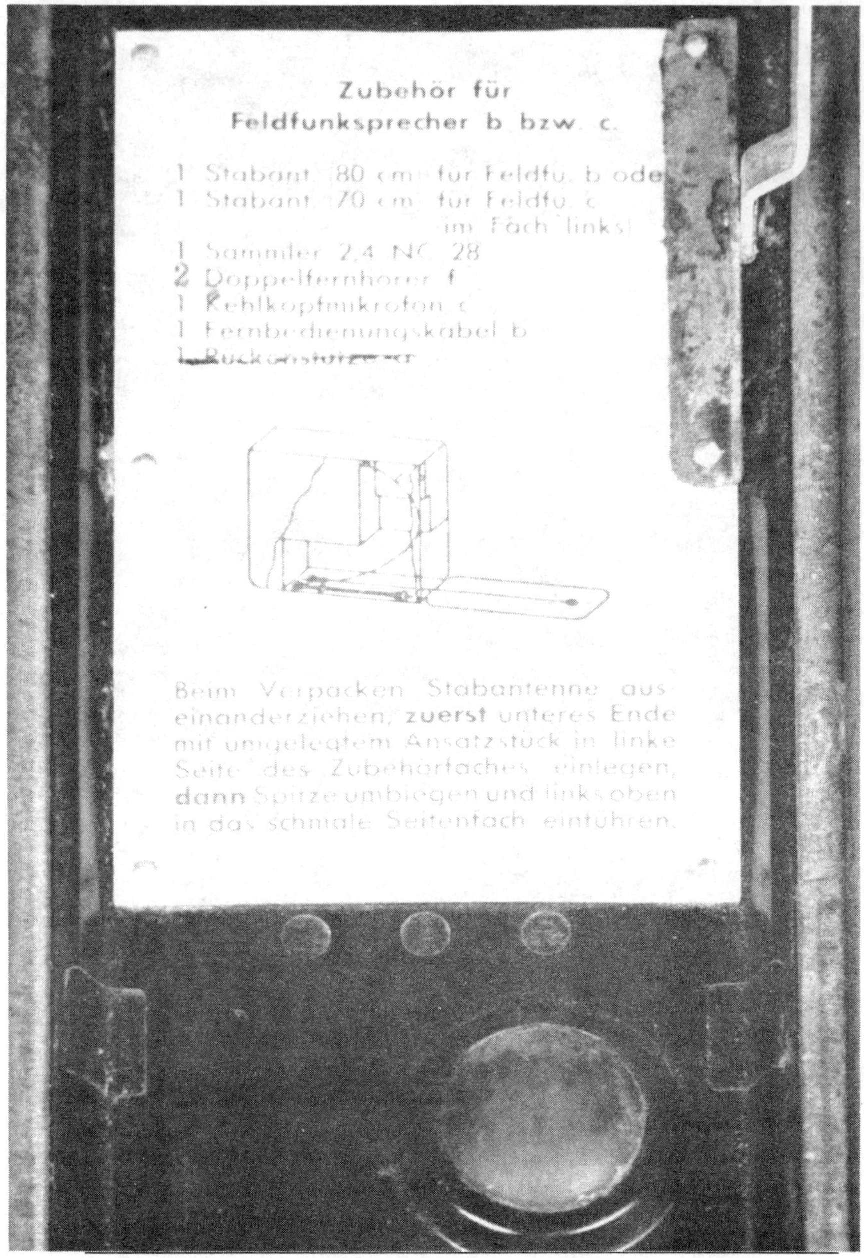

Figure 6. Data plate on the rear access panel. Note the listing of the contents and diagram for properly storing the antenna.

b. Uncover the calibration opening by turning the calibration-opening cover screw to the left (figure 4). The adjustment of this trimmer is possible only with the use of another set.

c. Set the **Frequenz-Einstellung** (tuning control) of both sets to the channel marked with red. Then set the **Empf.-Nachstimmung** (receiver fine tuning) control of the set requiring adjustment, so that the red dot is visible through the guard strap hole. This is the normal position of this control.

d. With the other set transmitting a modulated signal, set the **Ein Aus** (on off) switch to **Ein** (on), adjust the **Frequ.-Prufg.** (calibration trimmer) for maximum signal and close the cover of the calibration opening. The **Empf.-Nachstimmung** (receiver fine tuning) control can now be used to tune in any station close to the frequency to which the **Frequenz-Einstellung** (tuning control) is set.

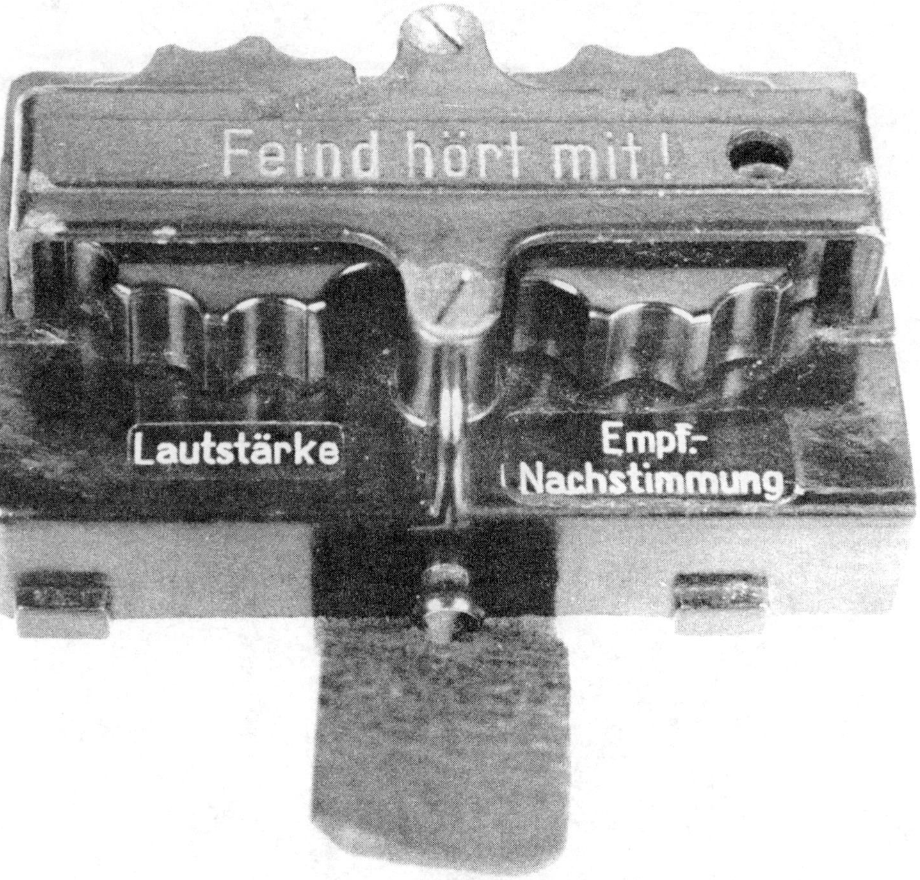

Figure 7. Close up of the remote control unit. Note the operational security warning label "Feind hort mit!" (The enemy is listening). The remote control unit can be worn on the operator's belt using the connecting cable, Fernbedienungskabel.

Figure 8. Rear view of the remote control unit showing the leather fastening strap.

Figure 9. Feld fu b. with cover removed and back rest (Rucken-stutze) installed.

7. BATTERY CHARGING: Charging requirements of nickel-iron-alkali batteries are different from the more common lead-acid batteries, and the following instructions should be observed.

a. Remove the two metal filler plugs before charging, and check the electrolyte level. If the plates are not covered, add enough distilled water to cover them.

CAUTION: NEVER PUT ACID IN THIS BATTERY

b. Charge the battery 10 hours at a 4.2-ampere rate, or 7 hours at a 6-ampere rate. The higher charging rate is preferable. The charge is complete when the voltage at the battery terminals measures 3.6 volts and remains at this value for 2 hours with the charger connected and delivering 6 amperes.

c. After the battery is charged, clean the filler plugs and replace them. Clean the top of the battery and the terminals, then go over the entire battery with a lintless oily cloth.

d. The battery will give better service if it is short-circuited and completely discharged just before charging. A hydrometer cannot be used to check the electrolyte, since its specific gravity remains practically constant during charge and discharge.

Figure 10. Rear view with antenna installed.

Figure 11. Close up view of the back rest support. The metal framework is made of sheet metal.

Figure 12. The inside of the cover. The instructions for operating the set are printed on the white plastic panel.

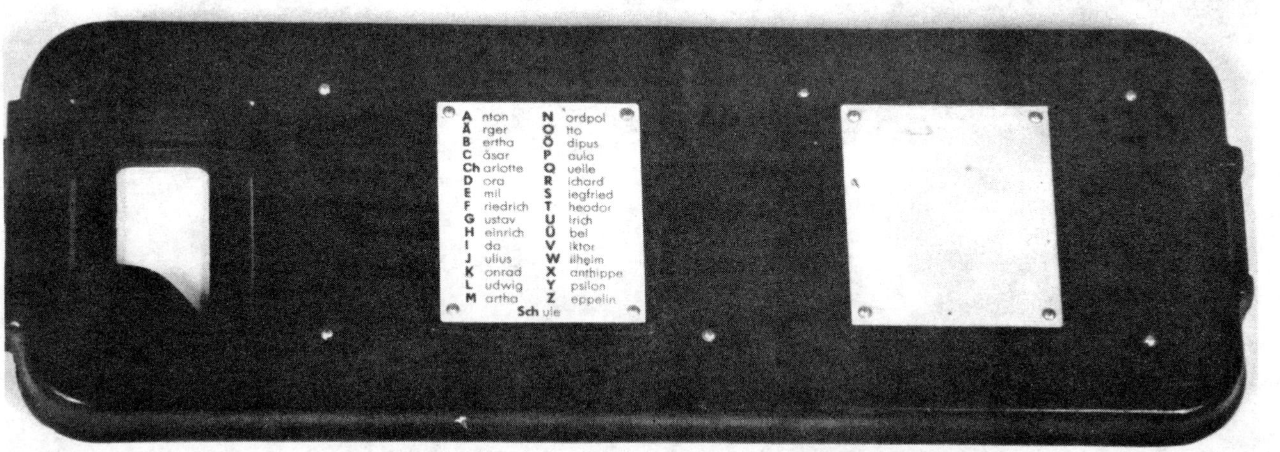

Figure 13. The exterior view of the cover. Note the writing block, phonetic alphabet plate, and access opening for the on-off switch.

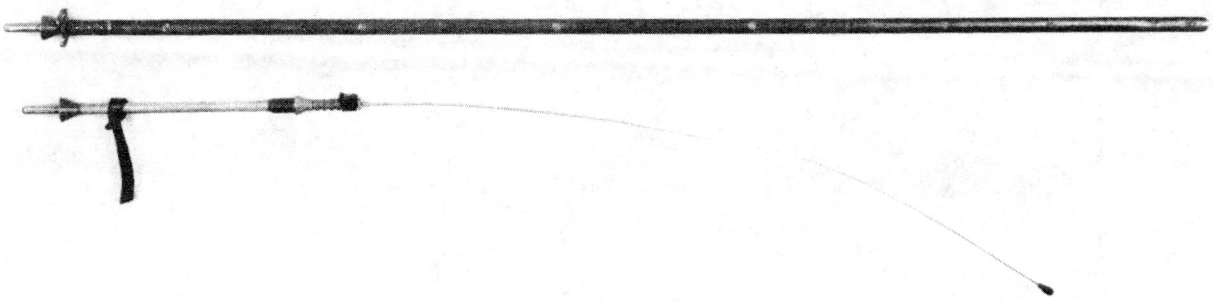

Figure 14. This photo shows two types of antenna issued with the set. The top antenna is much like a metal tape consisting of 6-7 sheets riveted together about every six inches or so. The antenna on the bottom is a two piece antenna consisting of the bottom or base and the top which is a metal wire about 3/16 inch thick. The base will be color coded to eliminate error in using a wrong antenna, red for feld fu b, green for feld fu c etc.

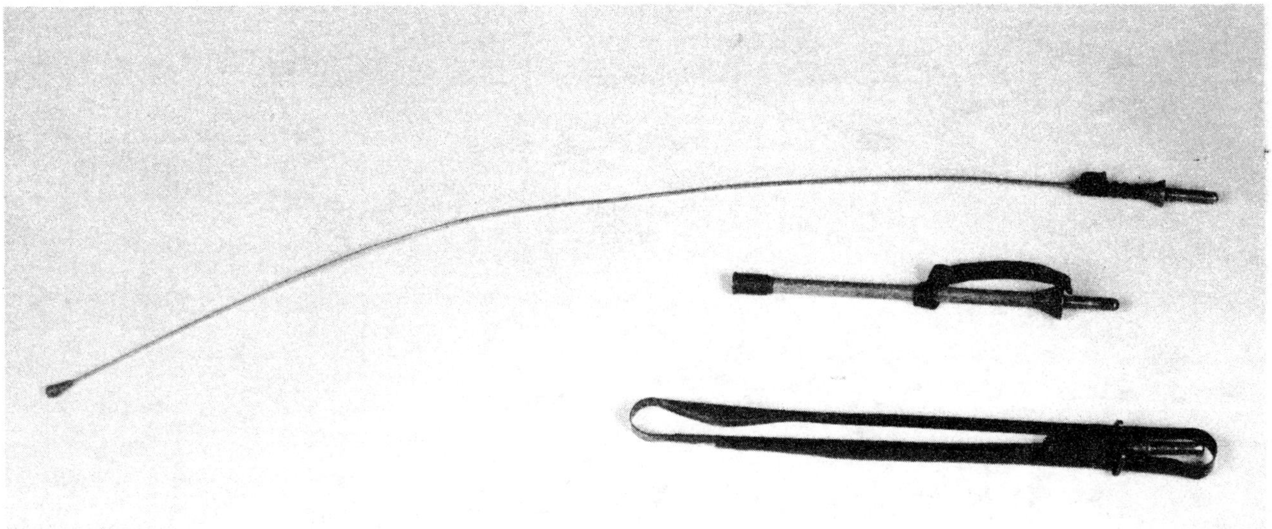

Figure 15. The antennas in storage configuration.

IV

The TORN. Fu. d2

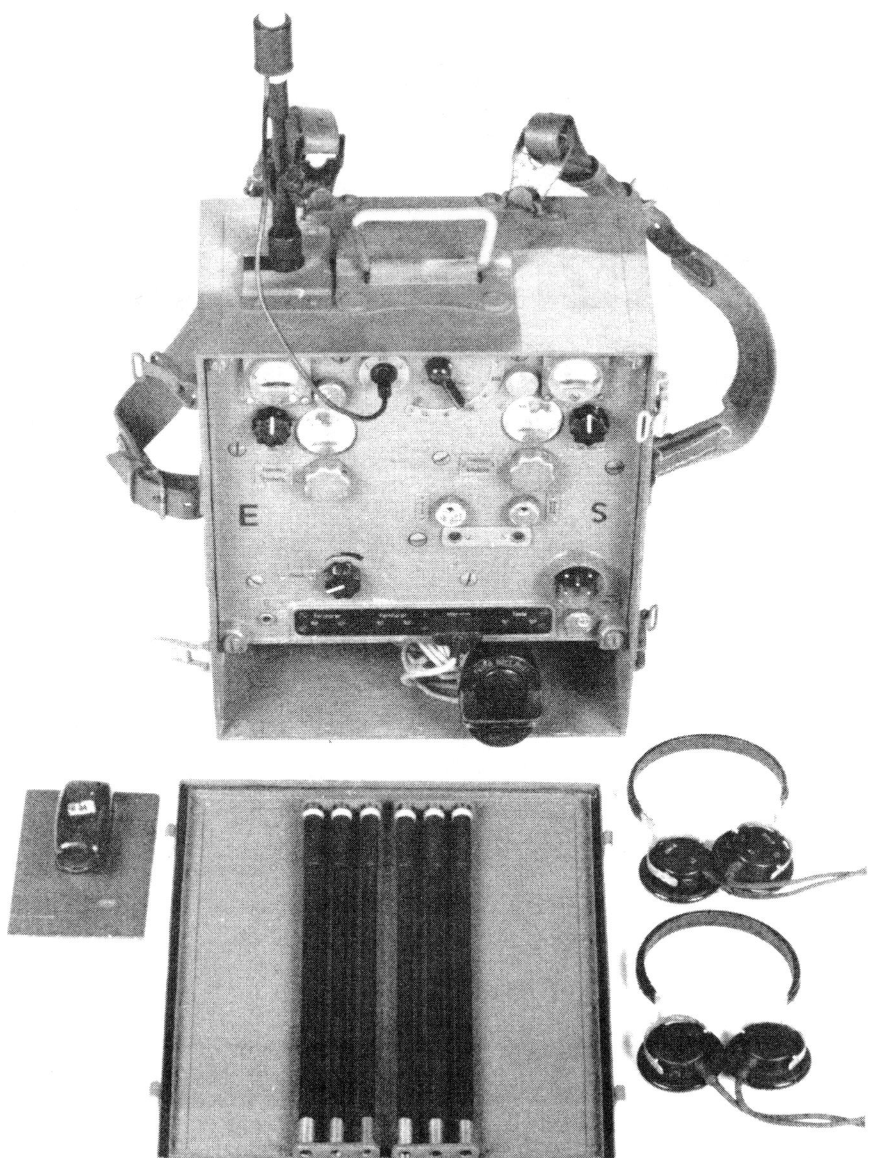

Figure 1. Accessories carried in the apparatus case and radio set storage compartment. The antenna mast sections are held in place in the inside of the front cover by spring fasteners. Not shown is the counterpoise cable. The counterpoise when attached to the set acts as an artificial radiation pattern which strengthens transmitting.

The TORN. Fu. d2

1. DESCRIPTION: The German radio set Tornister Funkgerat d2., abbreviated **Torn. Fu. d2.**, is "portable radio set d2." The set is a small, portable, high-frequency transmitter-receiver combination consisting of a three-stage transmitter and a six-tube superheterodyne receiver capable of operation on **Tg.**, Telegraphie (cw), or **Tn.**, Telephonie (voice). The complete radio set is contained in two packs, the transmitter-receiver unit being in the apparatus case, and the power supply and frequency calibrator in the accessories case. The accessories and spare parts are distributed between both packs. Figure 1 shows the accessories carried in the apparatus case. The complete list of the components is as follows:

1 - 2-volt storage cell, German type 2B38
2 - 90-volt plate batteries, German type DIN/VDE 1600
2 - Headsets
1 - Throat microphone
1 - Hand microphone
1 - Antenna mast
1 - Spare parts bag containing six tubes, German type RV2/800; two tubes, German type RL2Ts; and two microphone buttons
1 - Telegraph key
1 - Counterpoise consisting of a terminal plug and four rubber-covered rays about six feet in length
1 - Cord with three hooks which is used to mount a packcover above the operating panel in such a way as to provide additional protection against rain.
6 - Antenna sections which are carried in the place provided on the inside of the apparatus case cover.

2. PERFORMANCE DATA: The table below lists the performance data of the set.

PERFORMANCE DATA

Frequency range:	Transmitter	33.8 to 38.0 mc
	Receiver	33.8 to 38.0 mc
Types of signals emitted	Telegraphie, **Tg.** (cw), Telephonie, **Tn** (voice), amplitude-modulated	
Types of signals which may be received		cw, tone, and voice
Distance range	Cw	10 miles approximately
	Voice	2 miles approximately
Antenna type		Vertical, approximately 6 ft. in length
Type of transmitter		master oscillator-power amplifier (MOPA)
Number of tubes:	Cw	3
	Voice	4
	Master oscillator tube	type RV2P800
	Intermediate amplifier tube	type RV2P800
	Modulator tube	type RV2P800 (also used as a side tone oscillator and receiver a-f amplifier)
Power amplifier tube		type RL2T2

Number of dial divisions	100
Frequency separation between dial divisions	42 kc approximately
Method of calibration	fixed crystal oscillator (7,000 kc)
Power output	1 watt
Type of receiver	superheterodyne
Number of tubes	6
1	r-f amplifier type RV2P800
1	h-f oscillator type RV2P800
1	mixer type RV 2P800
1	i-f amplifier type RV2P800
1	2d detector type RV2P800
1	a-f amplifier type RV2P800
Intermediate frequency	2,100 kc approximately
H-f oscillator frequency	2,100 kc above incoming signal
Type of c-w oscillator	regenerative 2d detector
Sidetone in the set	yes-receiver a-f amplifier acts as audio oscillator to produce sidetone on cw
Zero beat transmitter to receiver	impossible
Power supply	batteries
Filament	2-volt storage cell
Plate	two 90-volt dry batteries
Grid bias for receiver	3 volts
Transmitter current drain when used in 2-way communication:	
Filament	2.2 amps approximately, on cw; 2.0 amps approximately, on voice
Plate	40 ma approximately, on cw 30 ma approximately, on voice
Receiver current drain:	
Filament	1.15 amps approximately
Plate	25 ma approximately

Note: When German batteries are used, the plate and grid use a supply which is tapped at 3 volts.

3. BATTERY INSTALLATION:

a. **General.** If the proper German batteries are available, they should be connected in the accessories case as shown in figure 2-①. The storage cell, German type 2B38, will fit into a compartment provided. Each of the two Anode Batteries (B batteries) also fits into a compartment of the case. Figure 3 shows compartments and placement of equipment in the accessories case.

b. **Filament Supply Substitutes.** When the German storage cell is unavailable, a 2-volt storage cell of any manufacture may be used. Dry cells of 3 volts may be used in place of the storage cell.

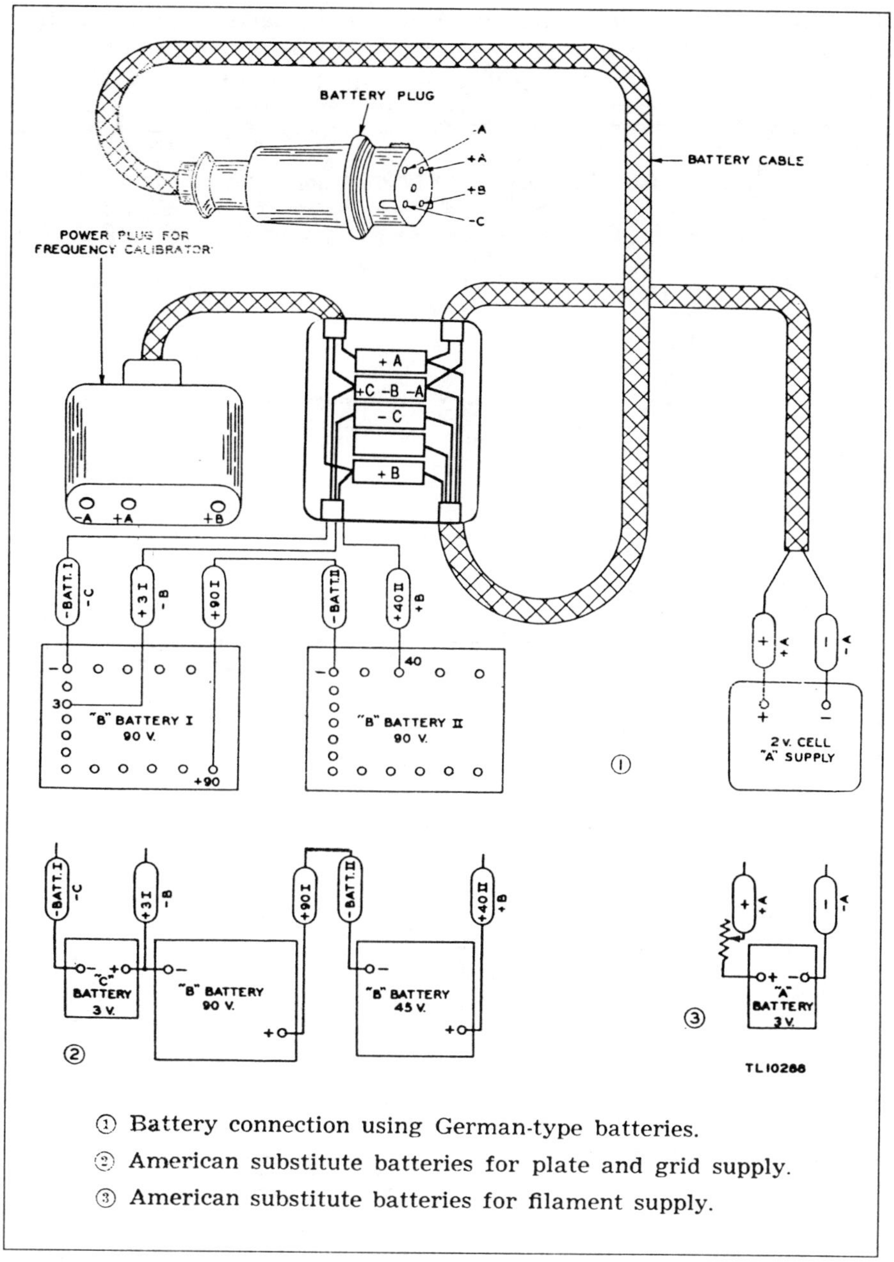

Figure 2. Power supply connections for German radio set Torn. Fu. d2.

However, a dropping resistor must be connected (fig. 2-③), and provisions made for adjusting the resistance in this circuit when changing from the **Empf.** (receive) position to either of the two transmit positions marked **Tg.** (cw) or **Tn.** (voice).

Note: Adjust the value of the dropping resistor so that the filament voltmeter reading does not exceed 2.2 in any position of the control switch (fig. 4). The value of this resistor is approximately 1/2 ohm on **Tg.** and **Tn.** and approximately 1 ohm on **Empf.** This adjustment will lengthen the life of the tubes by preventing the application of excessive voltage to the filaments of the tubes.

c. Plate and Grid Supply Substitutes. When German Anode Batteries (B batteries) are not available, B batteries of American manufacture may be substituted as shown in figure 2-②.

Note: A separate C battery must be connected for grid supply when American B batteries are used for plate supply.

d. **Battery Life.** The expected life of substitute American batteries is given in the table below. Other batteries which supply 2 volts for filament, 3 volts for the grid, and 235 volts for the plate may also be used. The batteries listed, however, will give reasonable life for a continuous filament drain of 2.2 amperes and a continuous plate drain of 40 milliamperes (continuous transmitting). Intermittent use or reduced drain such as in **Empf.** (receive) will, of course, greatly increase the life of the batteries. Since the drain on the C battery is very small, the expected life of these batteries is not given.

4. SET INSTALLATION:

a. Battery.

(1) Connect the batteries as shown in figure 2, and described in paragraph 3.

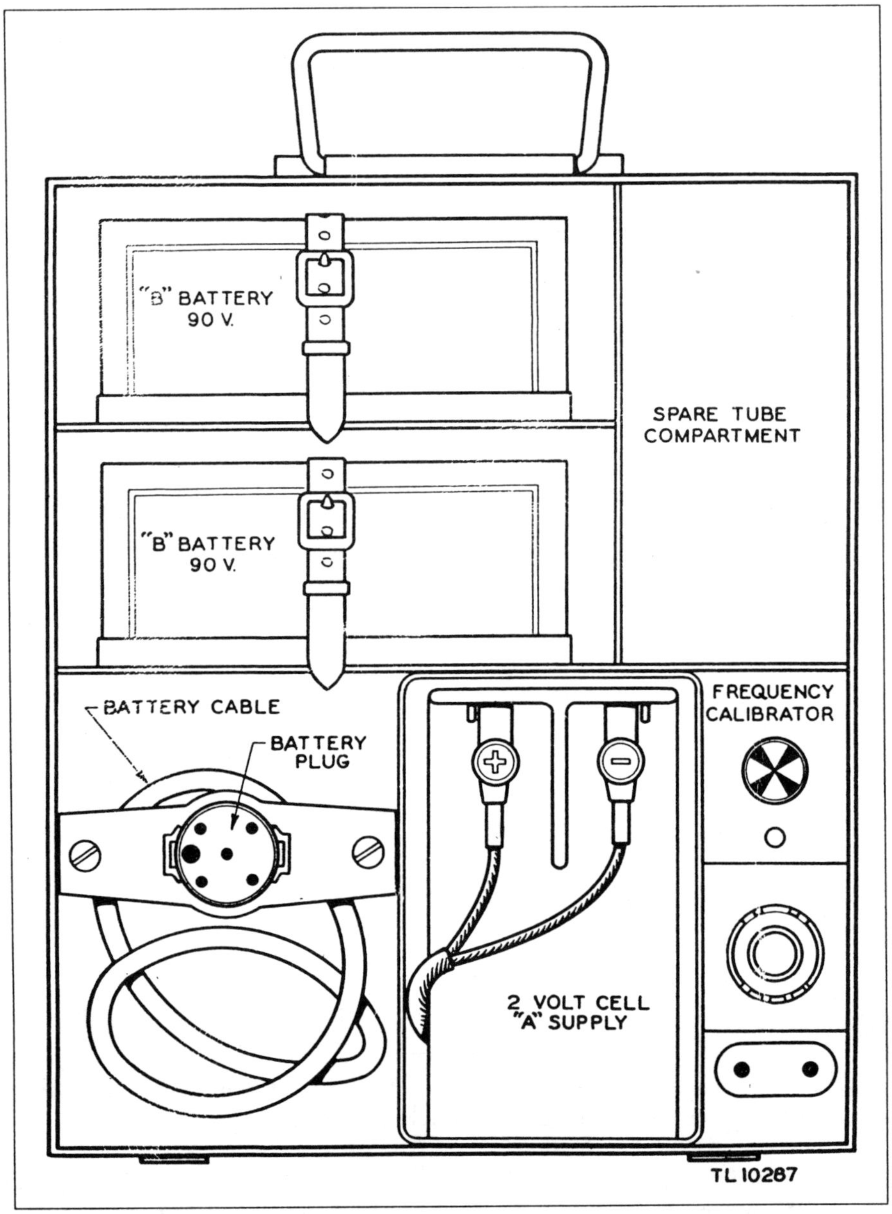

Figure 3. Placement of equipment in accessories case of German radio set Torn. Fu. d2.

(2) Connect the battery plug to the five-contact plug on the lower right of the set. Do not force this plug as it will only plug in one way. (Refer to figures 4 and 5 for the location of jacks and controls).

b. **Antenna.** Either the regular antenna or the remote antenna may be used. The remote antena affords operation with a distance of about 12 feet between the set and the antenna, which may be desirable when operating inside a tent, dugout, etc. The remote antenna equipment is carried in a separate waterproof bag. The regular antenna is easier to erect.

(1) REGULAR ANTENNA.

(a) Remove the antenna base and lead from the accessory shelf, and mount on the antenna mast bracket on top of the set. Tighten the large wingnut.

(b) Assemble the antenna (sections are in the apparatus case cover), and mount upright on the base already installed.

(c) The antenna lead connects to the jack marked **Ant**. (antenna) on the front panel of the set.

(d) Spread out the counterpoise cables on the ground and connect to the jack marked **G** (counterpoise) on the left lower corner of the front panel.

(2) REMOTE ANTENNA.

(a) Remove the various parts of the remote antenna from the waterproof bag.

(b) Mount the supporting mast in an upright position on a tent rod.

(c) Mount the antenna base to the supporting mast and tighten the wingout.

(d) Assemble the antenna (sections are in the set cover lid), and mount on the antenna base already installed.

(e) Brace the antenna structure with four guy ropes.

(f) Connect the shielded cable to the antenna and to the **Ant**. (antenna) jack on the set. About 12 feet of shielded cable is provided.

c. **Microphone.** Plug the microphone, either the throat or hand type, into the jack marked **Mikrofon** (microphone). Any single button carbon microphone equipped to plug into the jack will work satisfactorily.

d. **Telegraph Key.** Plug the telegraph key into the jack marked **Taste** (key). Any telegraph key will work.

e. **Headset.** Plug the headset into either of the jacks marked **Fernhorer** (headphones). Any high impedance headset is satisfactory. The set is now installed and ready for operation.

5. **OPERATION:** After the set is installed (pars. 3 and 4) it is put in operation as follows:

a. **Receiver Operation.**

(1) To receive, turn the main switch on the upper central part of the front panel to **Empf**. (receive).

(2) Read the battery voltages on the meter at the top left corner of the front panel. The filament voltage, shown as soon as the set is turned on, should be between 1.8 and 2.2 (indicated within the red section of the scale). Press the blue button marked 180V on the voltmeter to obtain the plate voltage. This should be between 130 and 140 (indicated within the blue section of the scale.)

(3) To tune the receiver, turn the control marked **Frequenz Einstellg**. (tuning control) directly under the dial on the **E** (receiver) side of the panel until the desired dial reading appears. The dial is calibrated from 0 to 100. The control marked **Fein** (fine) provides a fine adjustment of the h-f oscillator frequency which permits following a drifting signal and also changing the tone in the headset.

(4) When receiving voice or tone-modulated signals, keep the control marked **Ruckk**. (volume and regeneration) turned far enough to the left to prevent oscillation. When receiving c-w signals, this control may be turned to the right to produce the desired tone and the volume in the headset.

Figure 4. The Torn. Fu. d2. front view. The area in the bottom is used for storage.

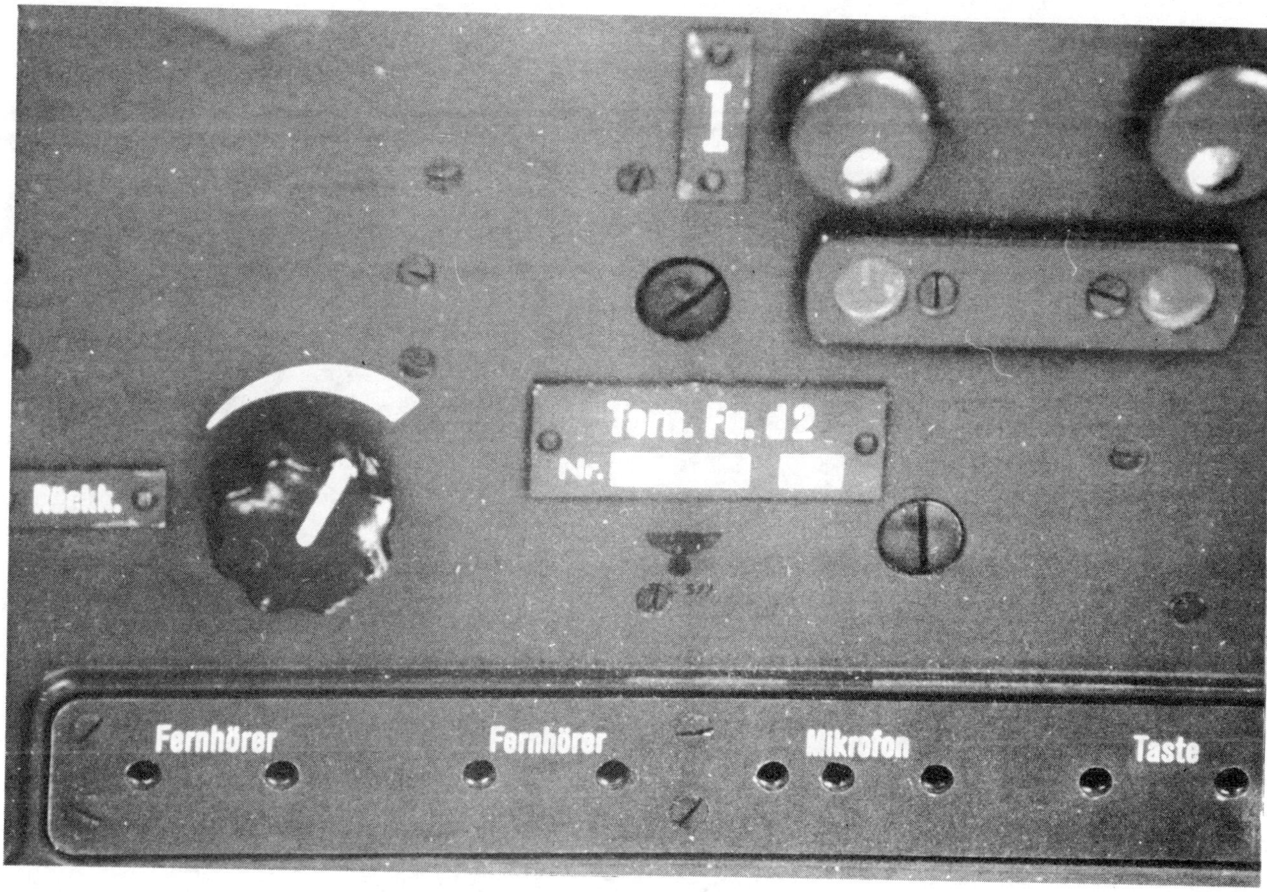

Figure 5. Close up view of the instrument panel from bottom to top: injacks for headphones, microphone and telegraph key; volume control; data plate; preset channel windows and locks. The ability to preset frequencies is a very important tactical consideration, especially during enemy contact and periods of darkness.

b. Transmitter Operation.

(1) To set the frequency of the transmitter, turn the control marked **Frequenz Einstellg**. (tuning control) directly under the dial on the **S** (transmitter) side of the panel until the dial is set at the desired reading. The dial is calibrated from 0 to 100. The frequency separation between dial division is approximately 42 kc.

(2) To preset channels proceed as follows:

(a) Turn the transmitter tuning control until the desired channel number, either **I** or **II**, appears in its window below the tuning control (figure 4). **I** appears in the preset channel window on the left and **II** in the one on the right.

(b) Turn the preset channel locks cover, which is found directly above the window showing either **I** or **II**, to the left to provide access to the channel lock screw.

(c) Turn the channel lock screw to the left so that when the tuning control is turned, the channel number selected remains in its window.

(d) Turn the transmitter tuning control to the desired frequency.

(e) Tighten channel lock screw by turning to the right.

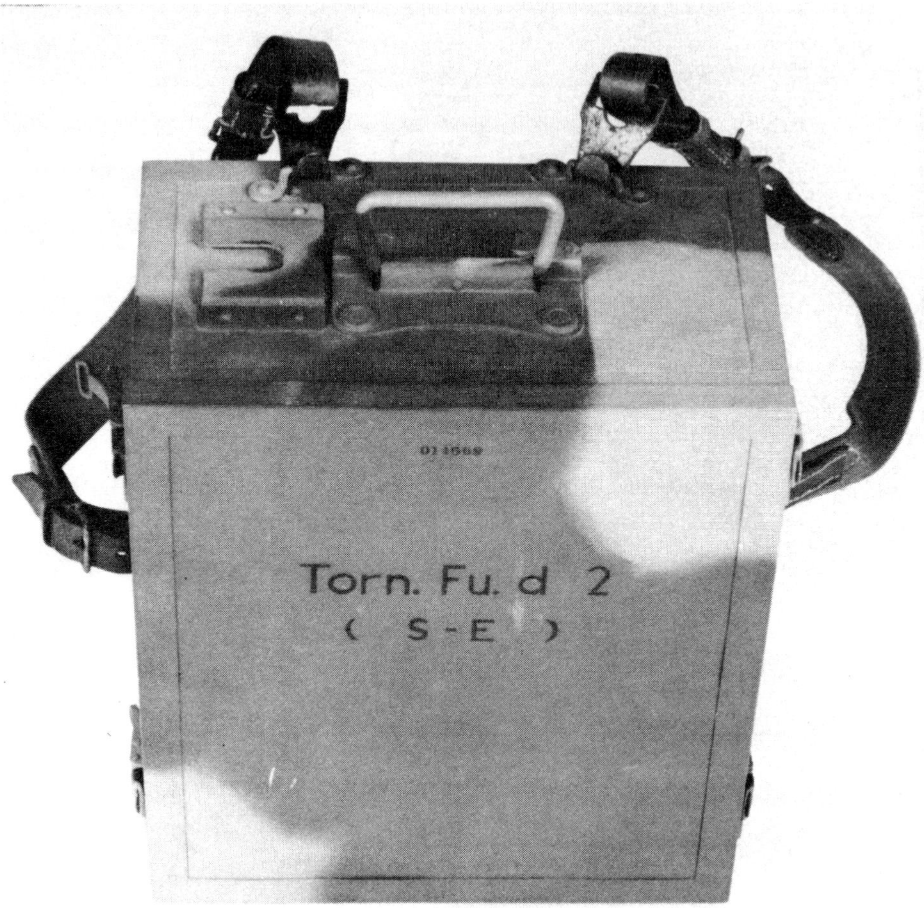

Figure 6. Front view of the Torn. Fu. d2. with cover in place. The serial number of the set was often stamped on the exterior panels for accountability purposes. Note the canvas carrying straps; these also exist in leather. Canvas straps gradually replaced leather as the war progressed.

(f) Whenever the channel number preset appears in its window, the transmitter is within 1 kc of the frequency orginally chosen.

(g) Set up the other preset channel in the same manner.

(h) The preset arrangement will also serve as a dial lock even though it is very easy to change frequency if desired.

(3) Turn the main switch to the **Tg**. (cw) position and check the battery voltages as instructed in paragraph 5 a (2). This is especially important when dry A batteries are used.

(4) Close the telegraph key and adjust the control marked **Ant. Abst**. (antenna tuning) for maximum reading on the antenna meter.

(5) To operate the transmitter on cw, use the telegraph key.

(6) To operate the transmitter on voice, place the main switch in the Tn. (voice) position and press the red knob on the microphone marked **Drucken** (press).

Note: When the main switch is in the **Tn**. (voice) position, the transmitter carrier is on.

(7) Due to the set design, it is impossible to zero-beat the transmitter to the receiver. The a-f amplifier tube of the receiver provides sidetone when operating on cw.

Figure 7. Rear view of the set. Again the serial number is present above the latch in the upper left section of the photo. Missing is the back support rest which slides into place by the use of metal guide grooves.

Figure 8. Rear view of the Torn. Fu. d2. removed from its case.

6. CALIBRATION: A crystal controlled frequency calibrator is located in the lower right side of the accessories case (figure 2). A power cable and plug is provided for connection to the batteries. The calibrator is placed in operation by pressing the push-button switch marked **Drucken** (press). Plate current flow is indicated by the appearance of white sectors in the instrument directly above the switch.

a. Calibration of the Receiver.

(1) Plug the headset into either of the jacks marked **Fernhorer** (headphones) on the front panel.

(2) Turn the main switch to **Empf.** (receive).

(3) Turn the control marked **Fein** (fine) until the stripe is vertical.

(4) Turn the control marked **Ruckk.** (volume and regeneration) to the right until oscillation starts. Normally a tone will be heard when the detector is oscillating.

(5) The red line near 28.2 on the receiver dial, corresponds to the frequency on which calibration is to be made.

(6) Press and hold the push-button switch marked **Drucken** (press) on the calibrator and using the control marked **Frequenz Einstellg.** (tuning control), tune the receiver to zero-beat.

(7) If the red pointer does not point to the red line on the dial, make the correction as follows:

(a) Open the receiver calibration adjustment cover located immediately to the right of the voltmeter.
(b) With a screwdriver, adjust the position of the red pointer so that it corresponds with the red line on the dial.

b. Calibration of the Transmitter.

(1) Plug the headset into the jack marked **Fernhorer** (headphones) on the frequency calibrator.

(2) Turn the main switch to **Tn.**(voice).

(3) Press and hold the push-button switch marked **Drucken** (press) on the calibrator and using the control marked **Frequenz Einstellg.** (tuning control), tune the transmitter to zero-beat.

Figure 9. Comparison view of the two sets. The set at left is finished in dark gray with white letters, while the set at the right is light olive with black letters. Note the camouflage paint scheme applied to the set at the right. The colors are sand, brown and green.

(4) If the red pointer does not point to the red line near 28.2 on the transmitter dial, make the following correction:

(a) Open the transmitter calibration adjustment cover located immediately to the left of the antenna meter.
(b) With a screwdriver, adjust the position of the red pointer so that it corresponds with the red line on the dial.

7. FIELD TELEPHONE OPERATION: It is also possible to modulate the transmitter over an ordinary field telephone line. Satisfactory operation may be obtained with about 1 1/4 miles of field wire. Open the cover on the jack, **z. Fspr**. (for connecting field telephone). The field wire line is connected to a plug which will fit into this jack and a field telephone is connected across the line at the transmitter. It is possible to converse with the operator at the remote location over the field wire line. It is necessary for the transmitter operator to set the main switch in the **Tn**. (voice) position to prepare the transmitter for voice operation. The remote operator will hear the receiver when the main switch is in the **Empf**. (receive) position.

Torn. Fu. d2.

Frequency table

Range: .... 33,800 Khz to 38,000 Khz

Dial increments: .... 0 - 100

1 increment .... = 42 Khz

0 - 33.800	30 - 35.060	55 - 36.110	80 - 37.160
5 - 34.010	35 - 35.270	60 - 36.320	85 - 37.370
10 - 34.220	40 - 35.480	65 - 36.530	90 - 37.580
15 - 34.430	45 - 35.690	70 - 36.740	95 - 37.790
20 - 34.640	50 - 35.900	75 - 36.950	100 - 38.000
25 - 34.850			

First figure is dial reading, second figure is frequency.

For each additional gradation, add 42.

Example: 6 = 34.052

Example: 72 = 36.824

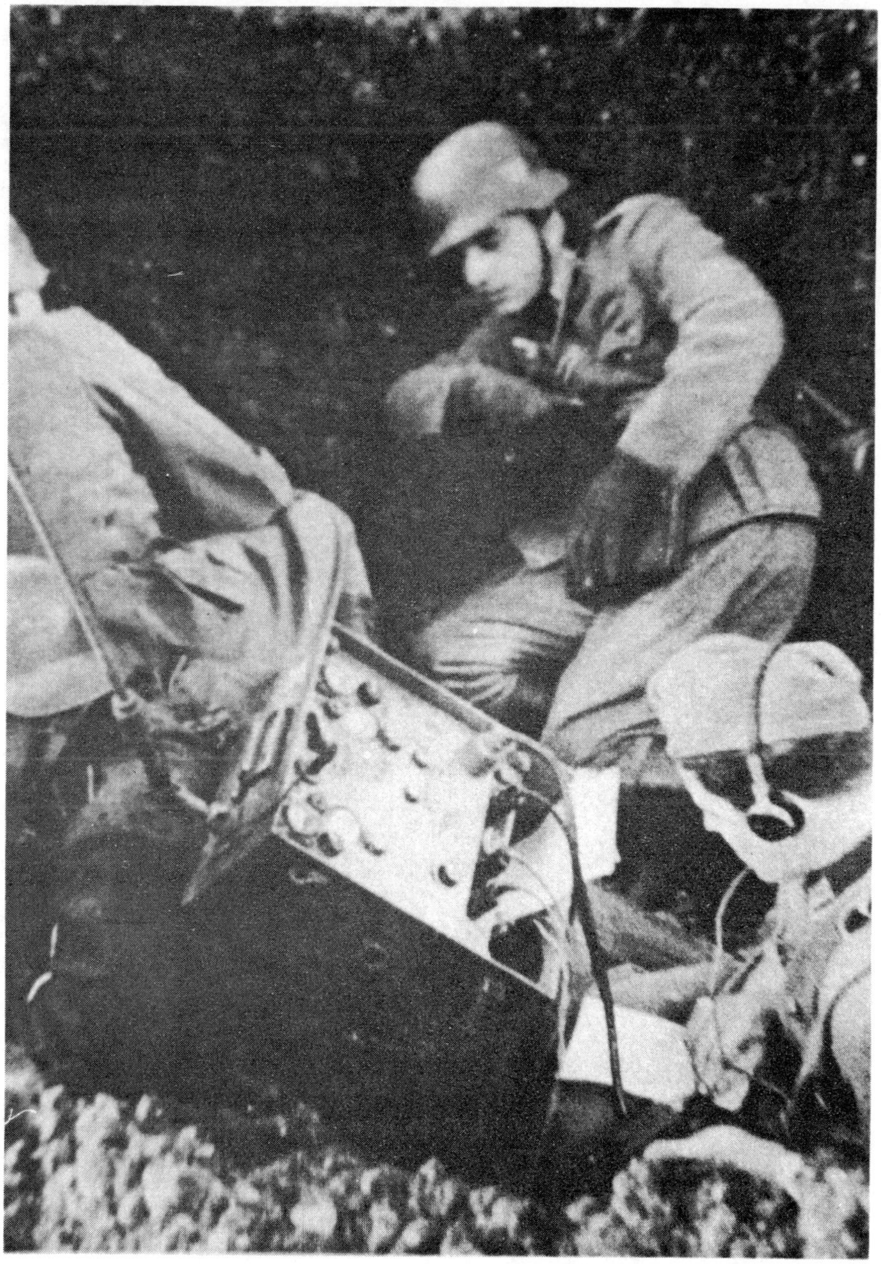

Figure 10. This radio team has taken cover in a bomb crater during the battle for France. The operator is sending a message using the telegraph key. Note the use of leather carrying straps and the position of the antenna.

Figure 11. This original photo shows the Torn. Fu. d2. being used on the march. The operator carries the accessory case and walks behind the assistant operator, who carries the actual radio.

V

The TORN. Fu. bl and f

This picture shows a Torn.fu.b1/f in operation during the battle for France. Note the backrest and antenna base.

The Torn. Fu. b1. and Fu. f.

1. DESCRIPTION:

a. The German radio sets Tornister Funkgerat b1 and f, Torn. Fu. b1.(portable radio set b1) and Torn. Fu. f. (portable radio set f) are identical except for a difference in transmitter frequency range. This set is a transmitter-receiver combination, capable of operation on **Tg**, Telegraphie (cw), or **Tn**,Telephonie (voice). The set is housed in two cases, one containing the transmitter and receiver; the other containing the power supply and accessories.

b. Performance data. The table below lists the performance data of the set.

PERFORMANCE DATA

Frequency range:	Receiver	3000 to 6600 kc approximately
	Transmitter Torn. Fu. b1.	3000 to 5000 kc approximately
	Transmitter Torn. Fu. f.	4500 to 6670 kc approximately
Type of signals emitted		Telegraphie, Tg (cw) and Telephonie, Tn (voice) amplitude modulated
Types of signals which may be received		cw, tone, and voice
Distance range:	Cw	12 miles approximately
	Voice	6 miles approximately
Type of transmitter:		master oscillator-power amplifier (MOPA)
Number of tubes:	Cw	2
	Voice	3
	Master oscillator tube	type RV2P800
	Power amplifier tube	type RL2P3
	Modulator tube	type RV2P800
	Method of calibration	quartz glow tube
	Power output	0.65 watts
Type of receiver:		superheterodyne
	Number of tubes	6
	1 r-f amplifier	type RV2P800
	1 h-f oscillator	type RV2P800
	1 mixer	type RV2P800
	1 i-f amplifier	type RV2P800
	1 2-d detector	type RV2P800
	1 a-f amplifier	type RV2P800
	Intermediate frequency	2000 kc
	H-f oscillator frequency	200 kc above incoming signal
	Type of c-w oscillator	regenerative 2d detector
Sidetone in set		none available
Zero beat transmitter to receiver		impossible
Antenna type		top-loaded rod or long wire

Power supply .... batteries
Filament .... 2-volt storage cell
Plate .... 125 1/2 volts (dry batteries)
Grid .... 4 1/2 volts (dry batteries)

NOTE: When German batteries are used, the plate and grid use a 130-volt supply which is tapped at 4 1/2 volts.

2. INSTALLATION AND OPERATION:

a. **Battery Installation.** If the proper German batteries are available, they should be connected in the accessories case as shown in figure 1. The storage cell, German type 2B38, will fit into a compartment provided. Each of the two **Anode Batteries** (B batteries) also fits into the compartment of the case.

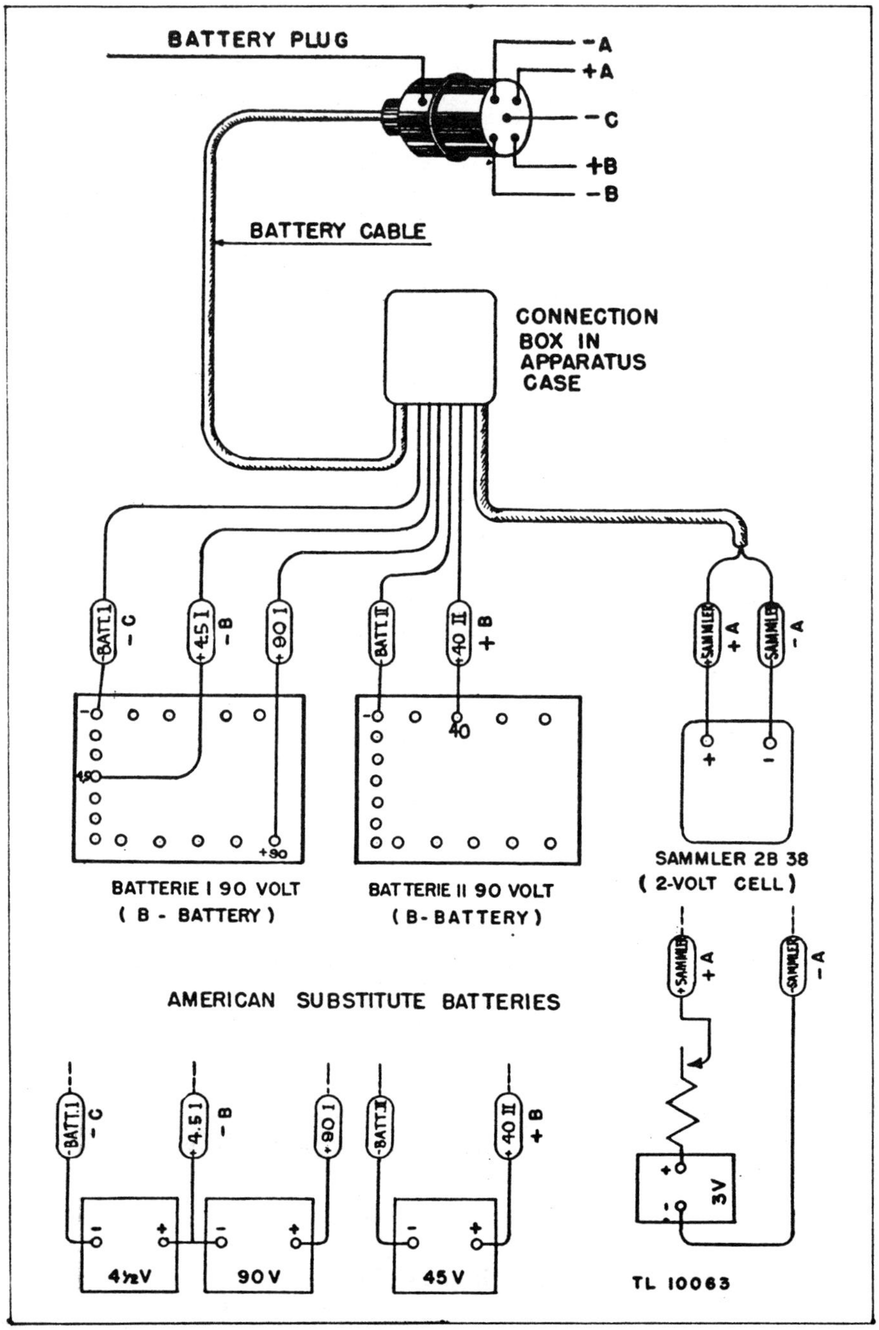

Figure 1. Battery wiring schematic.

b. **Filament supply substitutes.** When the German storage cell is unavailable, a 2-volt storage cell of any manufacture may be used. Dry cells of 3 volts may be used in place of the storage cell but a dropping resistor will have to be connected as shown in figure 1. Approximately 1 ohm will be needed in the **Empfang** (receive) position and approximately 0.4 ohms in the **Send. Empf.** (transmit-receive) position.

NOTE: Adjust the value of the dropping resistor, installed when dry cells are used for filament supply, so that the filament voltmeter on the set reads 2 volts. This will lengthen the life of the tubes by preventing the application of excessive voltage to the filaments in the tubes.

c. **Plate and grid supply substitutes.** When German **Anode Batteries** (B batteries) are not available, B batteries of American manufacture may be substituted as shown in figure 1.

NOTE: A separate C battery must be connected for grid supply when American B batteries are used for plate supply.

d. General set installation. The set has two types of antennas, high and low rod.

(1) The high-rod antenna will give a longer range than the low-rod antenna. If the rod antennas are not available, attach one end of a wire about 45 feet long to a tree, being sure to keep it insulated from the tree and not touching surrounding objects. Connect the other end of the wire to the set in the **Ant**. (antenna) jack.

(2) The low-rod antenna is made up of a vertical portion of three antenna sections and a top portion of eight antenna sections. (fig. 2).

(3) The high-rod antenna is made up of a vertical portion of seven antenna sections and a top portion of four antenna sections (fig. 2).

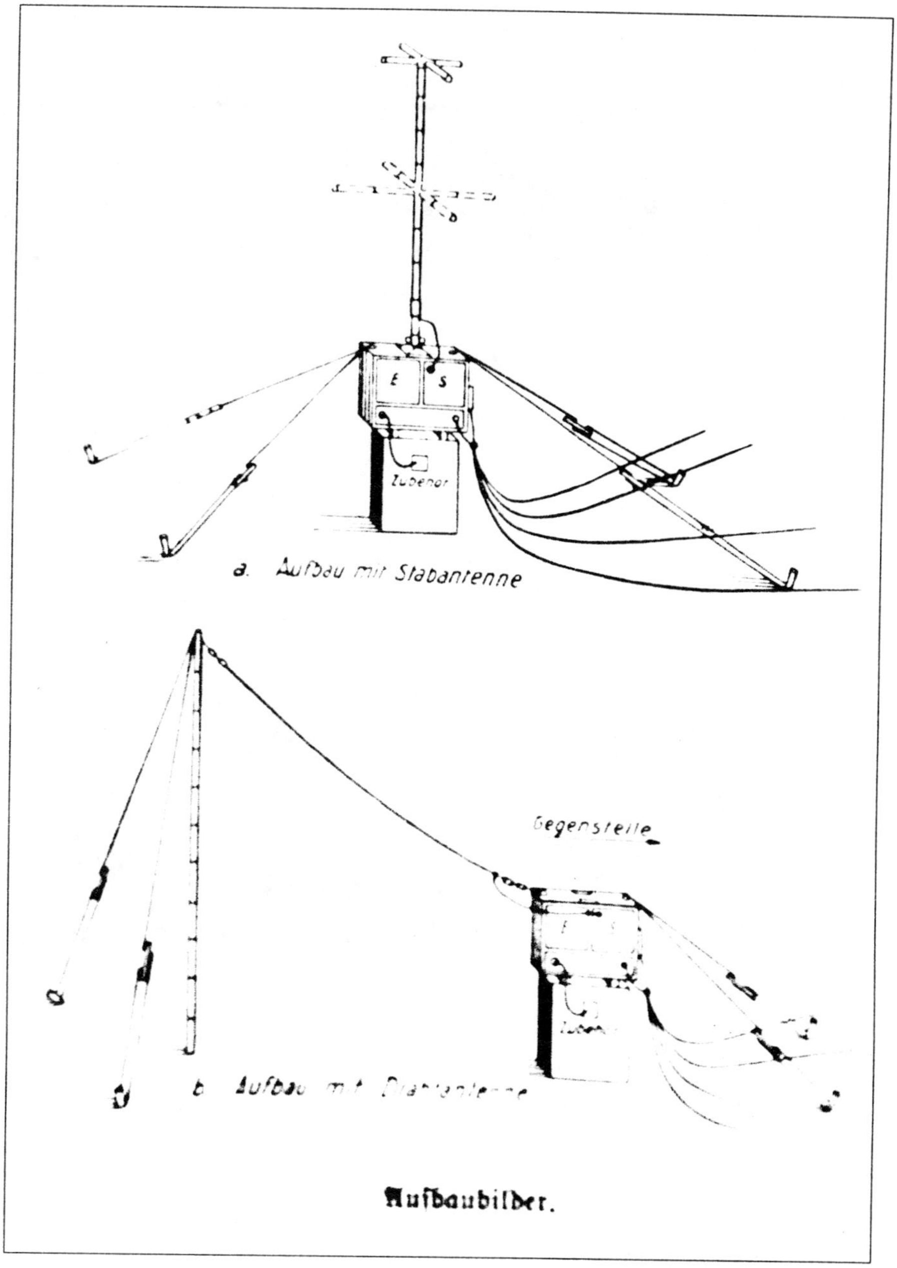

Figure 2. This illustration shows various ways of setting up the antenna (hi and low rod). In both cases the counterpoise cable has been set up to strengthen the transmitting capability.

(4) The battery plug is connected to the five-contact plug on the lower left of the set. Do not force this plug as it will only plug in one way.

(5) Plug the microphone, either the throat or hand type, into the jack marked **Mikrofon** (microphone). Any single-button carbon microphone equipped to plug into the jack will work satisfactorily.

(6) Plug the telegraph key into the jack marked **Taste** (key). Any telegraph key will work.

(7) Plug the headset into any of the jacks marked **Fernhorer** (headphones). Any high impedance headset is satisfactory.

(8) The set is now ready for operation. See figures 3-5 for location of jacks and controls.

Vorderansicht des Apparatetornisters.

Figure 3. Photograph taken from the original manual D 940/3. The letters correspond to the following components.

a. *voltmeter*
b. *retaining screw*
c. *handle*
d. *volume control*
e. *receiver tuning control*
f. *receiver dial*
g. *antenna jack*
h. *antenna meter*
i. *antenna switch*
k. *antenna tuning control*
l. *transmitter tuning control*
m. *transmitter dial*
n. *preset channel lock*
o. *preset channel lock*
p. *retaining screw*
q. *handle*
r. *battery plug*
s. *headphone jacks*
t. *retaining screw*
u. *power switch*
v. *microphone jack*
w. *jack for connecting field telephone*
x. *telegraph key jack.*

e. **Operation of the set.** After the set is installed, as described above, it is put into operation as follows:

(1) To receive only, turn the power switch on the lower portion to **Empfang** (receive).

(2) To transmit or receive, turn the power switch to the **Send. Empf.** (transmit-receive) position.

(3) To set the frequency of the receiver, turn the control marked **Frequenz Einstellung** (tuning control) on the **E, Empfanger** (receiver), panel of the set until the desired dial reading appears. Frequency in kc=dial readings x 10.

(4) To receive c-w signals, keep the control marked **Tn-Lautstarke-Tg** (volume) to the right of vertical toward **Tg** (cw).

(5) To receive voice or tone-modulated signals keep the volume control to the left of vertical toward **Tn** (voice).

Figure 4. Close up view right side of the set, transmitter section.

(6) To set the frequency of the transmitter, turn the control marked **Frequenz Einstellung** (tuning control) on the **S, Sender** (transmitter), panel of the set until the dial is set at the desired reading. Frequency in kc=dial readings x 10.

(7) To preset channels proceed as follows:

(a) Turn the transmitter tuning control until the desired channel number, either **I** or **II**, appears in its window below the dial. **I** appears in the preset channel window on the left and **II** in the one on the right.

(b) Turn the preset channel lock marked **Fest Lose** (locked unlocked) above the window so that **Lose** (unlocked) is uppermost.

(c) Turn the transmitter tuning control to the desired frequency.

(d) Turn the preset channel lock so that **Fest** (locked) is uppermost.

(e) Now, whenever the channel number preset appears in its window, the transmitter is within 1 kc. of the frequency originally chosen.

(f) The other preset channel may be set up in the same manner.

(g) The preset arrangement will also serve as a dial lock even though it is very easy to change frequency if desired.

(8) The switch marked **Stab-Ant. Draht-Ant.** (rod-antenna wire-antenna) changes the antenna circuit for the use of either type antenna. Place in the **Stab-Ant.** position for a high- or low-rod antenna and in the **Draht-Ant.** position for a wire antenna.

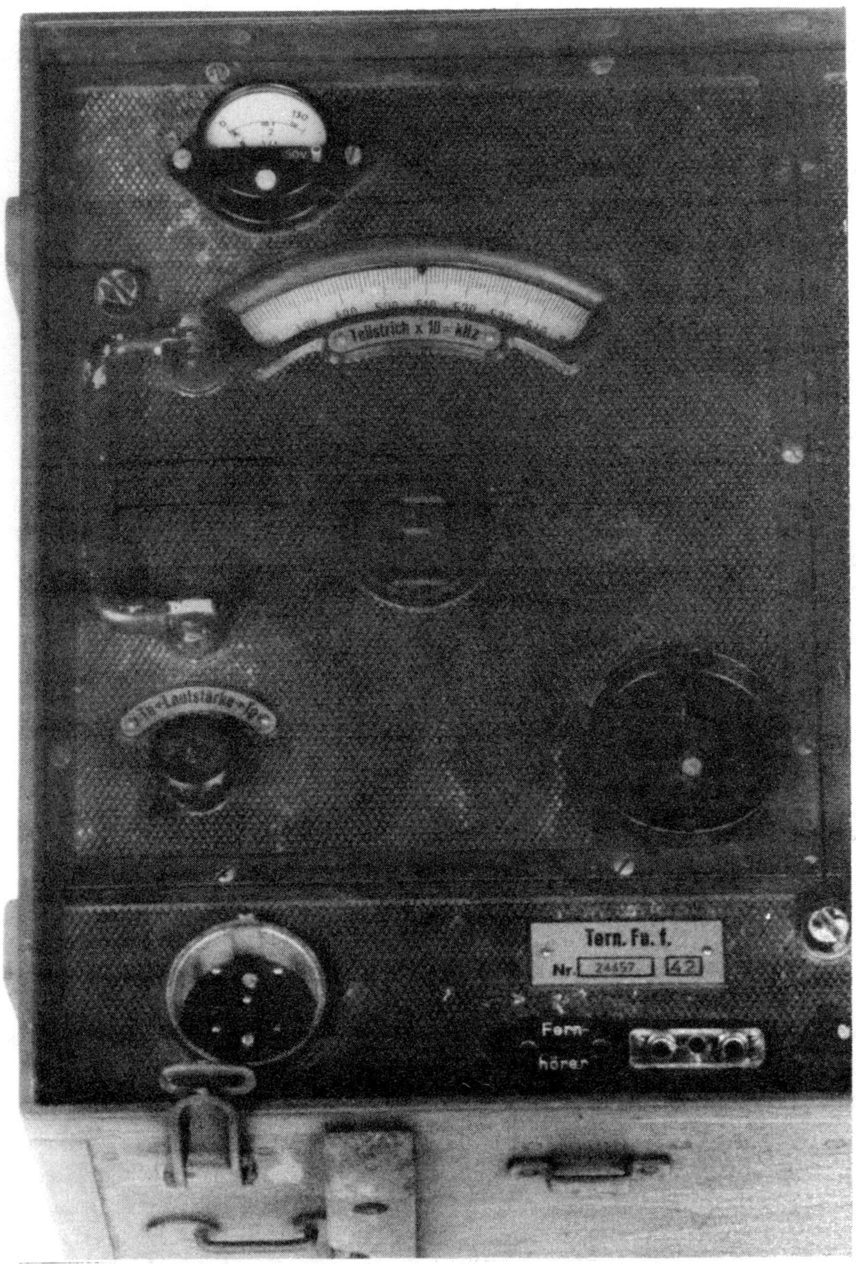

Figure 5. Close up view left side of the set, receiver section.

(9) With the power switch in the **Send. Empf.** position, press the key and adjust the control marked **Ant.-Abstimmung** (antenna tuning) until the meter marked **mA** (antenna current) reads maximum. The transmitter is now tuned and ready for operation.

(10) To check the filament cell or to adjust the resistor when substitute dry cells are used, turn the power switch to either **Empfang** or **Send. Empf.** and read the meter marked **V** (volts). It should read 2 in the red portion of the scale. When substitute dry cells are used, it will be necessary to readjust the resistor at the supply when switching from **Empfang** to **Send. Empf.** as indicated in paragraph 2b.

(11) To check the plate supply, turn the power switch to **Empfang** or **Send. Empf.** and push the blue button on the meter case marked **150V.** The meter should read 130 in the blue portion of

the scale. If the meter reads over 130 but less than full scale, the plate batteries are good.

(12) To operate the transmitter on cw, use the telegraph key; to operate the transmitter on voice, merely press the microphone switch. The necessary circuit changes are taken care of by the relays in the set.

(13) Due to the set design, it is not possible to zero-beat the transmitter to the receiver or to obtain sidetone in the headset.

3. MAINTENANCE: Detailed maintenance instructions on this set are not included. Simple operating precautions should be observed when set fails to operate, as follows:

a. Check the condition of the batteries with the voltmeter on the panel.

b. Check all cords, plugs, and connections. Much of the trouble will be in cording and connections.

c. Replace the tubes in order, starting with the antenna and progressing toward the master oscillator in the transmitter. In the receiver, start with the a-f amplifier and work to the r-f amplifier. To change the tubes it is necessary to remove the set from its case. Do this by loosening the three retaining screws which are on the front panel of the set (fig. 3). Then lift the set from the case by the two handles on the front. Figure 6 shows the rear of the set after it has been removed from its case and the position of all of the tubes.

NOTE: The i-f amplifier of the receiver and the modulator of the transmitter is the same tube; therefore, if the set is satisfactory in one or the other of its functions, the trouble is probably not in this tube.

Figure 6. The set removed from case, rear view.

d. If the simple procedures outlined above do not make the set operate, send it back to a signal depot. The components may be used to repair other sets. WE CAN USE THE GERMAN PARTS TO FIX OUR OWN OR OTHER GERMAN SETS.

4. **CALIBRATION:** In order to calibrate the transmitter, it is necessary to remove the set from its case.

a. Set up the transmitter for operation on cw.

b. Set the transmitter dial to the red line, located approximately at dial reading 468.

c. Press the key and look in either of the openings marked **Kontrollquarz** (calibration crystal). One of these openings is on the back and the other at the top of the set. If a glow is seen in the tube, the set is correctly calibrated. If there is no glow, insert a screwdriver in the opening marked **Frequenzkorrektur Sender** (for adjusting transmitter calibration) and adjust the screw until the glow is observed. The transmitter dial is now assumed to be correct. Calibration of the receiver should not be attempted by the operator.

5. **REMOTE OPERATION:**

a. **General.** The set includes a case which contains the following equipment for remote voice operation (fig. 7).

(1) Reel containing 50 meters (165 feet) of cable.

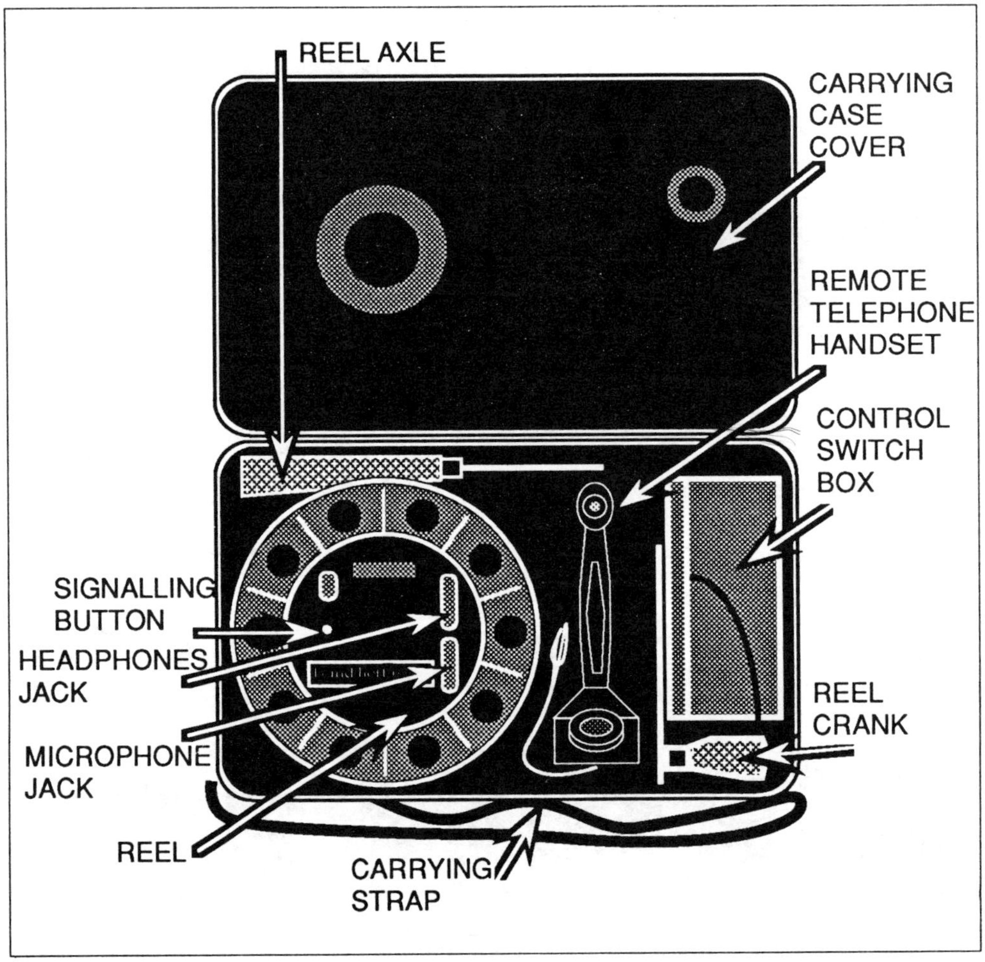

Figure 7. This line drawing shows the remote operation equipment.

(2) Reel axle for winding and unwinding reel.

(3) Reel crank used in conjunction with the reel axle.

(4) Control switch box used at the transmitter location.

(5) Remote telephone handset.

b. **Installation**.

(1) The control switch box is located at the transmitter and is connected to the five-contact jack on the set marked **Fernbesprechung** (remote control jack) by the five-contact plug on the switch box.

(2) Plug the microphone (hand or throat type) into the jack marked **Mikrofon** (microphone) on the switch box.

(3) Plug the headset into the jack marked **Fernhorer** (headphones) on the switch box.

(4) Plug the cable from the reel into the five-contact plug provided on the control switch box.

(5) Using the reel axle and crank, unwind the cable on the reel and place the reel at the remote location.

(6) Plug the three-contact plug of the remote telephone handset into the jack on the reel drum marked **Mikrofon** (microphone).

(7) Plug the two-contact plug of the remote telephone handset into the jack on the reel-drum marked **Fernhorer** (headphones).

NOTE: A microphone and headset may be used instead of the remote telephone handset as long as they are plugged into the correct jacks. The remote telephone handset may be used at the transmitter instead of the microphone and headset.

(8) The installation is now complete and ready for operation.

c. **Operation.**

(1) Set up the equipment as indicated in paragraphs 2e and 5b.

(2) The control switch box is equipped with a switch which selects any one of the three types of operation (fig. 8).

(3) For operation of the transmitter from the transmitter location, place the selector switch of the control switch box on **Ortsbespr**. (local)

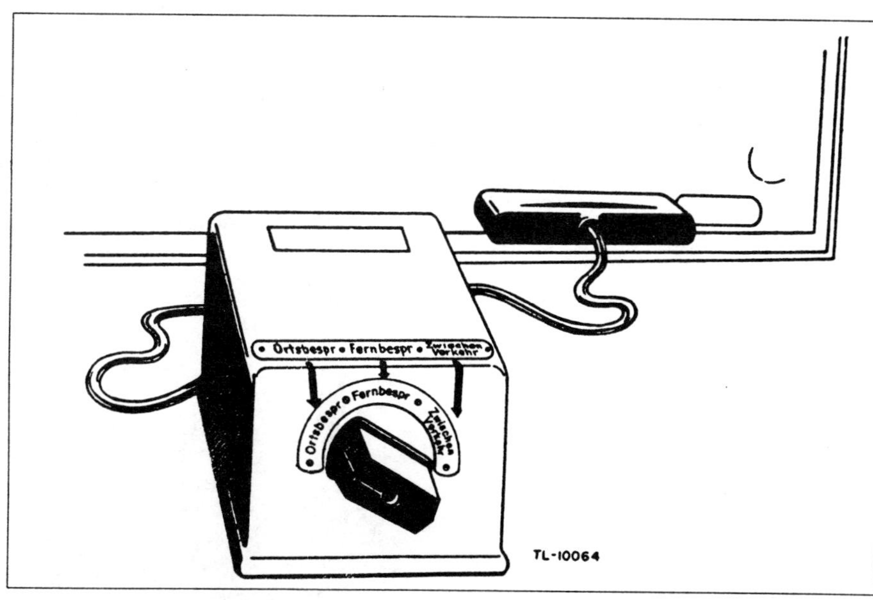

Figure 8. This line drawing shows the control switch box for remote operation.

position. In this position, the local transmitter operator puts the transmitter carrier on when he presses his microphone switch. Either the local or remote operator may modulate the transmitter when the carrier is on and either operator may hear the other operator's conversation. Both operators hear the receiver when the transmitter is off.

(4) For operation of the transmitter from the remote location, place the selector switch on the control switch box on **Fernbespr.** (remote) position. In this position, the remote operator puts the transmitter carrier on when he presses his microphone switch. Either operator may hear the other operator's conversation. Both operators hear the receiver when the transmitter is off.

(5) When it is desired to have the transmitter operator and the remote operator converse over the cable without putting the transmitter carrier on, place the selector switch on the control switch box in the **Zwischen Verkehr** (telephone) position. Operation of either microphone switch does not put the transmitter carrier on. Both operators hear the receiver at all times in this position.

6. FIELD TELEPHONE OPERATION: It is also possible to modulate the transmitter over an ordinary field telephone line. Satisfactory operation may be obtained over about 1 1/2 miles of field wire. The line is connected to a plug which will fit into the jack on the set marked **Zur AnschluBklinke f. Feldfernsprecher** (for connecting the field telephone) and a field telephone is connected across the line at the transmitter. It is possible to converse with the operator at the remote location over the field wire line. It is also possible for the remote operator to modulate the transmitter provided the transmitter operator has pressed his microphone switch to put the transmitter carrier on and prepare it for modulation. The remote operator will hear the receiver when the transmitter is off.

Figure 9. The Torn. fu. f, case with the cover on. The actual radio illustrated is a Torn. fu. f. The cover went to another set.

Figure 10. Close up view of the 4 section antenna socket.

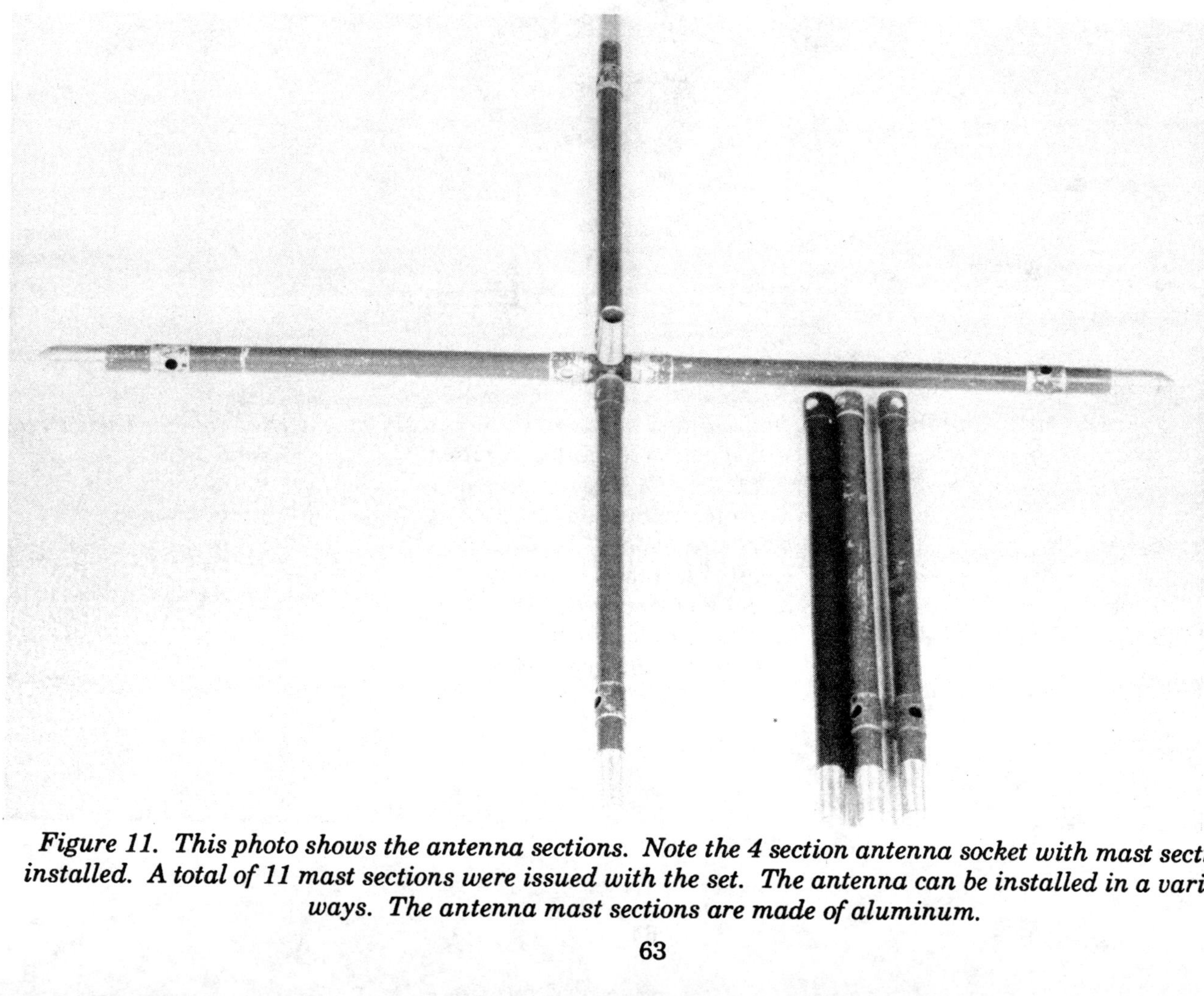

Figure 11. This photo shows the antenna sections. Note the 4 section antenna socket with mast sections installed. A total of 11 mast sections were issued with the set. The antenna can be installed in a variety of ways. The antenna mast sections are made of aluminum.

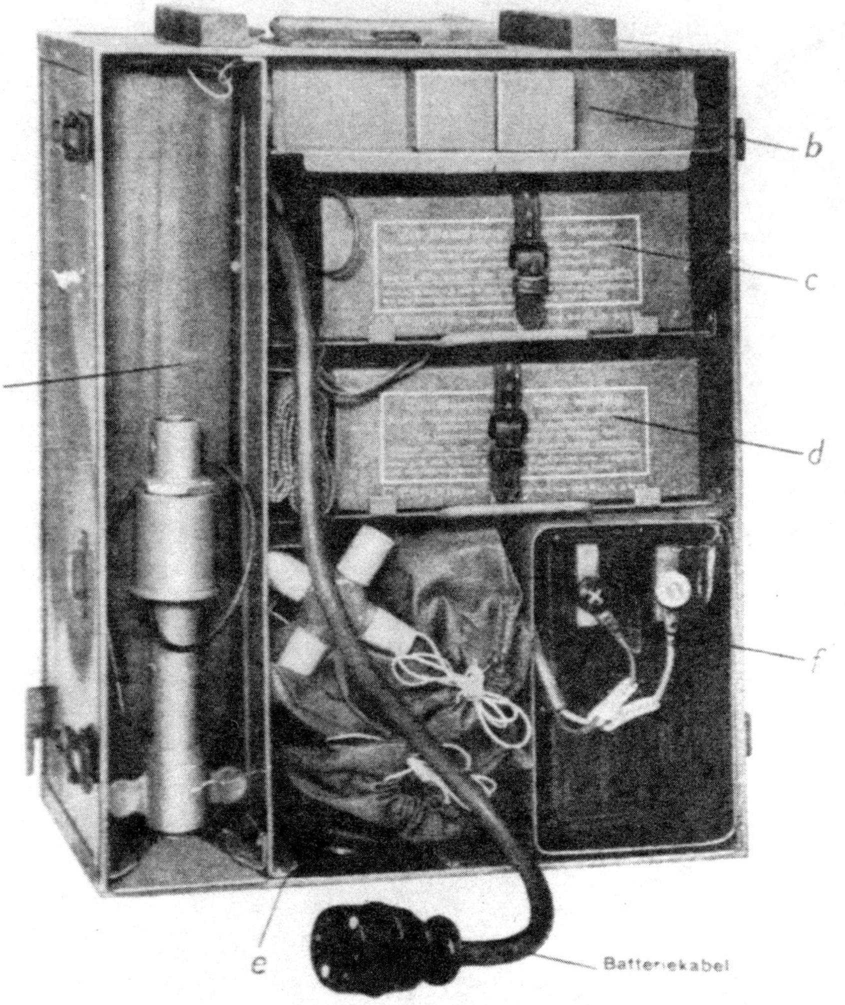

Vorderansicht des Zubehörtornisters.

Figure 12. An excerpt from an original manual showing the accessory case with contents. The letters correspond to the following items:

a. Antenna Sections
b. Spare Tubes
c. 90 Volt Anode Batteries
d. 90 Volt Anode Batteries
e. Counterpoise in Bag
f. Storage Cell Type 2B38

Figure 13. Rear view of the Torn. Fu. b1 accessory case. This case is painted in the three color camouflage scheme (sand, brown, and green). Note the back rest.

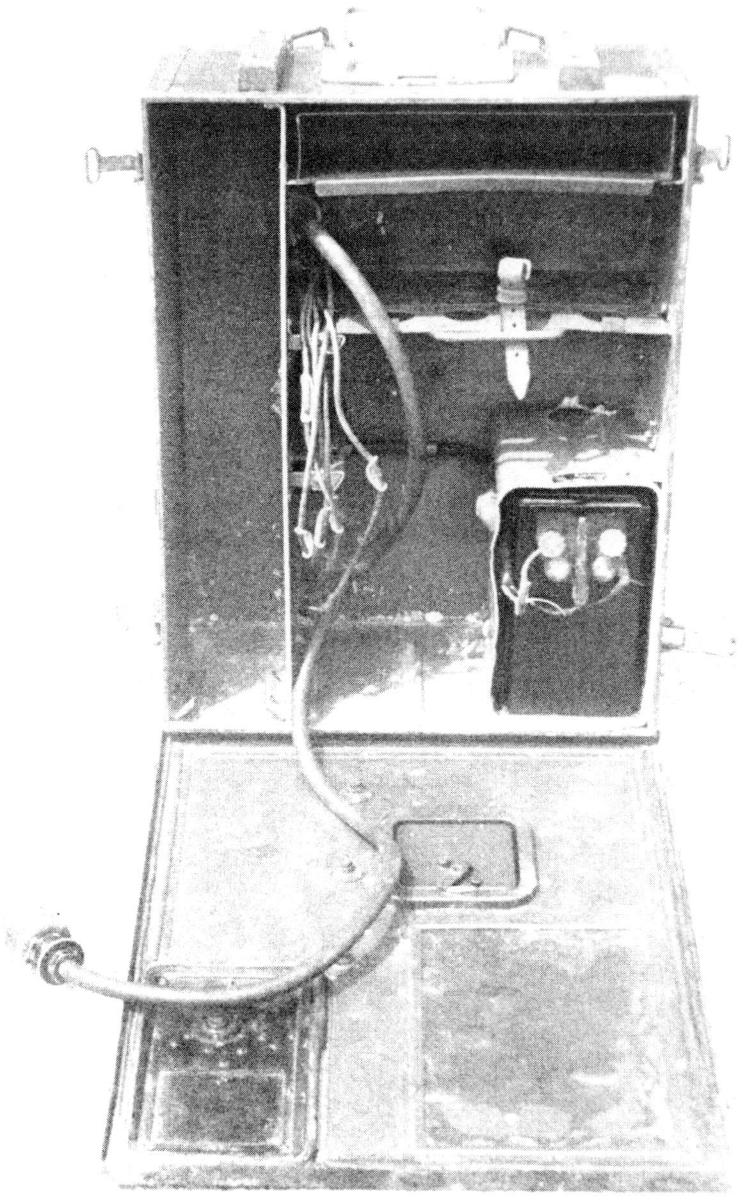

Figure 14. Interior view of the accessory case. The lower battery shelf is missing.

Figure 15. Front view of the Torn Fu. b1 Zubehör (accessories) case. Note the cable access door.

Figure 16. Another view of the accessory case with the battery cable door in the opened position.

Figure 17. The storage battery 2.4 NC 58z another type of battery used with the Torn Fu. b1/f series radios.

Figure 18. The Torn. fu.F front view. The actual finish is field gray. The labeled plates are in light green finish. Note the post war repair on the headphone and telegraph jacks (clear plastic).

Figure 19. Instructions for use manual D 940/3 1937.

Figure 20. This signal team has set up in a stationary position and is transmitting. The operator at the right is using the handset that is designed to be used when wearing the steel helmet.

Figure 21. The torn fu b1 shown in this photo has been installed on the sidecar using an angle iron bracket. The bracket is probably a field modification constructed by the maintenance company.

VI

The Kleinfunksprecher d

Figure 1. The kleinfunksprecher d.

The Kleinfunksprecher d

1. DESCRIPTION: The Kleinfunksprecher "d" is a highly compact small portable infantry transceiver capable of operation on Tn (Voice) communication. It can be used in networks with the German radio set Torn Fu. d2 the Torn Fu. f and the 10 watt sender. The set consists of 2 parts: the apparatus case, worn on the operator's left shoulder strap, and the battery box worn on the operator's belt.

The following items make up the set: one 1.6 meter bond antenna, the receiver/transmitter, battery box, hand or throat microphone, headphones, transport case (Tpka), and instruction booklet. D1037/5.

2. PERFORMANCE DATA:

Frequency range .... 32.0 - 38.0 MHZ (7.9 - 9.4 meters)
Calibrated in channels and kilocycles

Range

Between two Small Wireless Sets 'd', or between
Small Wireless Set 'd' and 'Torn Fu d 2 .... 2 - 4 kilometers

Between Small Wireless Set 'd' and '10 W S c' using
'UKW E e' .... 2 -4 kilometers (while moving)

Between Small Wireless Set 'd' and 'Feldfu f' .... 1-2 kilometers.

Output .... approx. .2 watts

Modes

radiotelephony in simplex mode. Can be used both stationary and on the move. Point-to-point traffic; net operation.

Power supply

Dry batteries .... 1.4-volt filament battery (LS 1.4 bp)
150-volt anode (B-) battery (LS 150 bp)

Duty cycle .... 25 hours in normal operation
(20% transmitting, 80% receiving).

Antenna .... 1.6-meter band antenna (for custom-made wire antennas

Vacuum tubes .... 2 ea 'R L 1 P 2'
1 ea 'D D D 25'

Dimensions .... Carrying box (gross)
45 by 22.5 by 13.5 centimeters.
Transceiver .... 13 by 7 by 20 centimeters
Battery box .... 11 by 10 by 7 centimeters.

Weight .... Carrying box (gross)
7.3 kilograms
Transceiver .... 1.6 kilograms
Battery box: .... 1.5 kilograms
Set (operational): .... 3.5 kilograms

3. **OPERATION:**

a. **General.**

(1) Open the transport box and remove the contacts.

(2) Install the headphones and throat microphone into the input jacks on the instrument panel as shown in figure 2a and b.

(3) Place the antenna into the antenna socket.

Figure 2a. Front view of the instrument panel from top to bottom: data plate, power switch, frequency dial, microphone jack (left) and headphone power cable jack (right).

(4) Open the battery box by unscrewing the two thumbscrews on the top of the box, (see figure 3). Insert and connect the anode and filament battery (figure 4). Close the box and retighten the thumbscrews. The box is now ready for use.

(5) Connect the power cord to the input jack on the instrument panel.

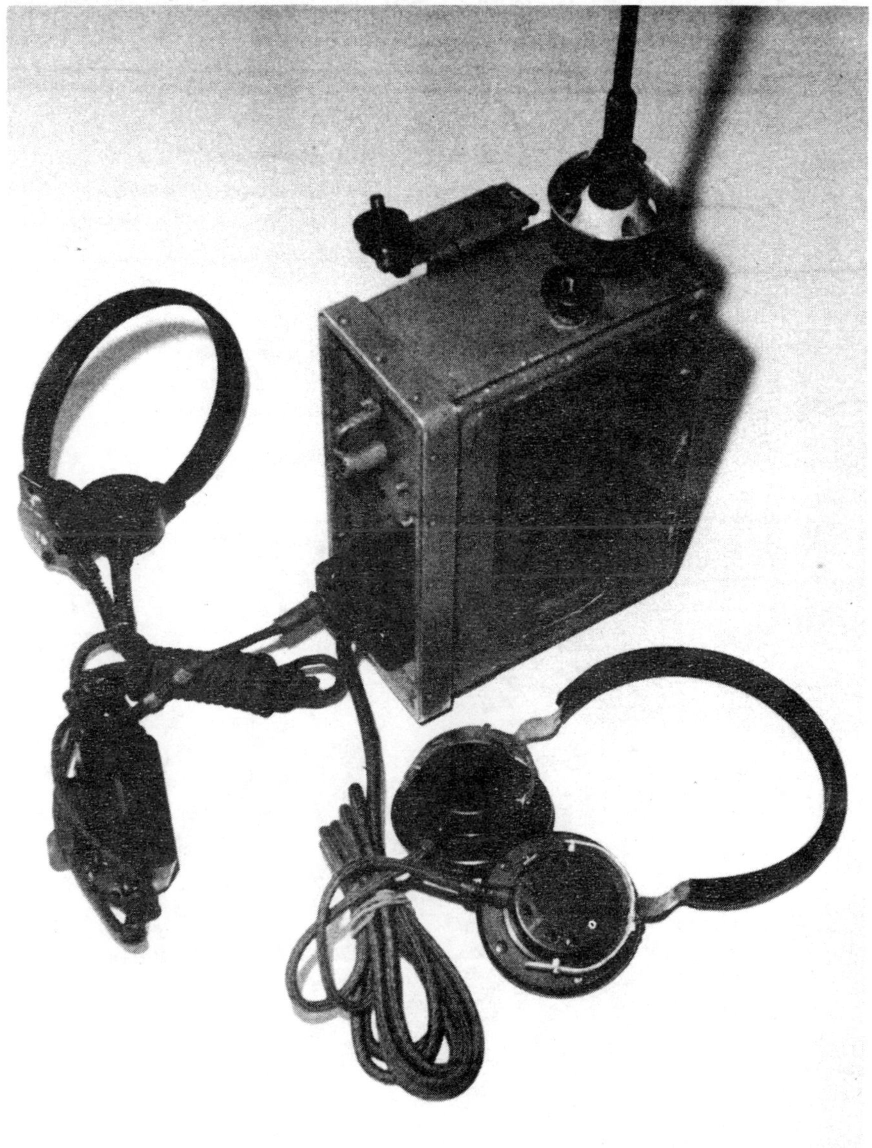

Figure 2b. The set with headphones and throat microphone installed.

b. Use in stationary deployment.

(1) Put the receiver/transmitter and battery box together by using the guide groves found on the back of the battery box and receiver/transmitter unit, (figure 5). To carry the set, slip the leather strap on the battery box into the bar bracket located on the top of the receiver/transmitter.

(2) Attach the antenna.

(3) Connect the power cord.

c. **Use on the March.**

(1) Attach the receiver/transmitter to the Y-strap support harness over the left chest.

(2) Attach the hook on the back side of the set to the web harness.

(3) Fasten the battery box to the belt using the leather strap. The battery box should be placed over the left hip to allow enough slack for the power cable.

(4) Follow steps as described in sub-paragraph (2) above.

(5) For use in a prone position, take the hook off the Y-strap and put the receiver/transmitter upright on the ground in front of the prone operator. When using the headphone headstrap (e.g. when wearing the steel helmet) remove the earpieces with connecting cables from the headphone set and attach them to the headstrap.

4. **ESTABLISHING AND CONDUCTING WIRELESS COMMUNICATIONS:** The set is operational if a noise is heard when switching-on the receiver, and listening-in is possible when talking into the microphone in the transmitting mode.

Figure 3. The battery box closed position. Note the leather carrying strap and guide grooves for attaching to the transceiver.

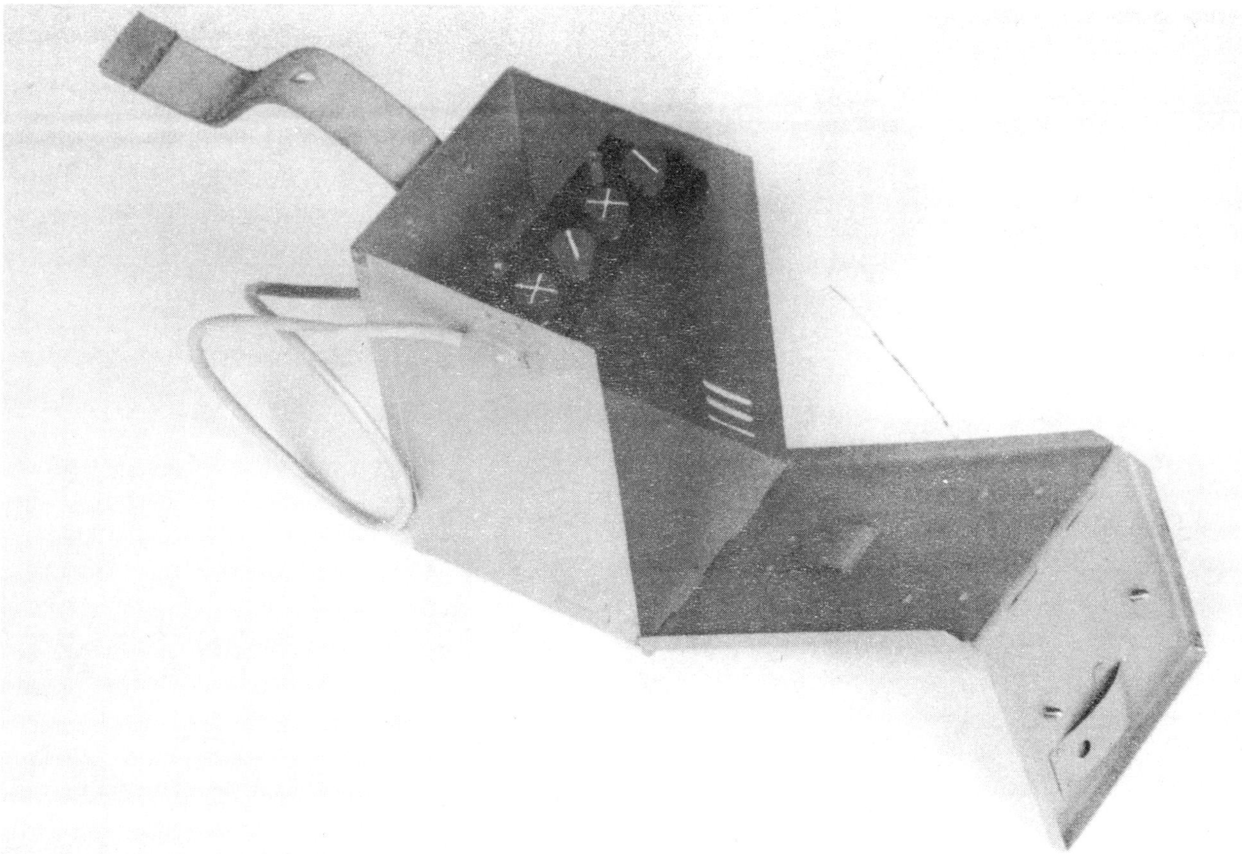

Figure 4. The battery box in the opened position.

a. Adjusting the frequency.

(1) Center the arresting spring (below tuning control) between the two stops. Press down spring. Turn tuning control knob to set the frequency desired. Black digits are for channels covered by (portable radio set) 'Torn Fu d2'. Red digits x 100 for kilocycles covered by (radio sets) '10 W S c' and 'Feldfu f', see figure 6.

(2) Turning of the tuning control to the left or right beyond the stops requires bypassing of the tuning lock.

b. Calling the far station. Set mode switch to "**Senden**" (transmitting). Call far station. (IMPORTANT NOTE: The position of the microphone switch is irrelevant. Setting the mode switch to "Senden" automatically readies the set for speaking/transmitting).

c. Receiving the far station. When the far station is a small wireless set, do not change the frequency setting as the far station will use the same setting. If a call is expected from a kleinfunksprecher d and the call is to be received on a portable radio set ('Torn Fu d 2') or a ' 10 W S c' or a 'Feldfu f', tune-in the frequency (as explained above) and search for the far station between the two stops.

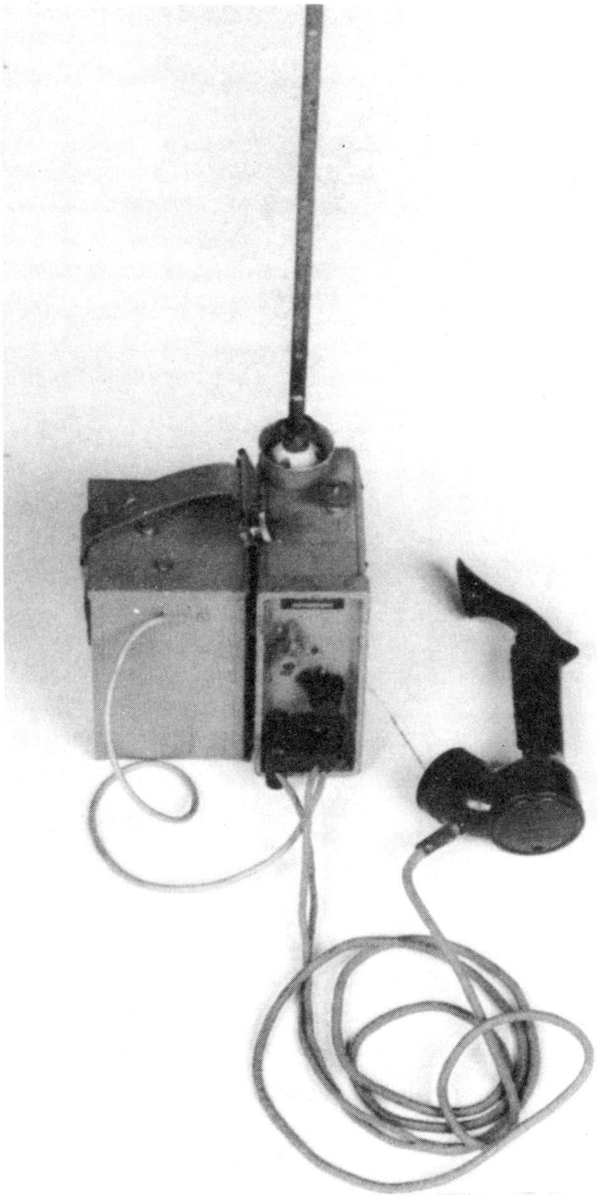

Figure 5. The Kleinfunksprecher d for use in stationary operation. The handset is depicted in this illustration.

Figure 6. This illustration shows the set removed from the case. Note the two rows of numbers.

d. Operation.

(1) Conduct radio traffic strictly in accordance with standard operational procedure.

(2) Do not change the frequency setting during radio traffic.

(3) Conserve battery power by keeping transmissions short.

(4) Be aware of the peculiarities of ultrashort-wave (VHF) communications.

(5) In radio communications with the portable radio set 'Tornisterfunkgerat (Torn Fu) d 2' as the far station, the operator of the 'Torn Fu d 2' must be informed that he must use both the tuning control knob and the receiver fine-tuning control to home-in on the frequency used by the kleinfunksprecher d when establishing the link.

Figure 7. Another view of the set removed from the case, opposite view from figure 6.

CAUTION! The receiver of the kleinfunksprecher d jams the receiver of the 'Torn Fu d 2' and that of the 'U K W E c' at distances up to 200 meters when using the same frequency or a frequency up to 5 channels, or 200 kilocycles, adjacent to it.

5. MAINTENANCE:

Troubleshooting Table

Malfunction	Test result (using testing device d/k)	Cause	To eliminate malfunction:
1. Transmitting or receiving not possible at all (no audible noise when listening or speaking)	Not enough voltage in battery.	Bad battery.	Replace battery.
	Instrument deflects left during battery test.	Wrong battery polarity.	Connect to correct terminals.
	No filament/anode current - or - filament anode current too high (short circuit).	Defective battery cable or connector	Replace (Whole battery box, if necessary).
	No filament/anode current.	All vacuum tubes defective.	Replace
	Filament/anode current too low.	Tube DDD 25 defective.	Replace
2. Transmission impossible:	Transmitter gauge does not light, but anode/ filament current is OK.	No contact from antenna spring	Bend antenna spring slightly upward.
a. no transmitter power.	Transmitter gauge does not light; anode/filament current too low.	Tube RL1P2 (tube no. 1) or DDD25 is defective.	Replace
b. no modulation	Transmitter gauge does not shine brighter when whistling into microphone; no listening-in tone.	'Kmf' or tube DDD25 defective.	Replace
3. Receiving impossible:			
a. no noise	Filament/anode current too low.	DDD25 defective	Replace.
b. noise too low	Filament/anode current too low.	'Dfh' defective.	Replace
c. no sensitivity.	Filament/anode current OK.	No contact from antenna spring.	Bend Antenna Spring slightly upward
After tube replacement (always if DDD 25 is replaced!), a frequency test should be made.			

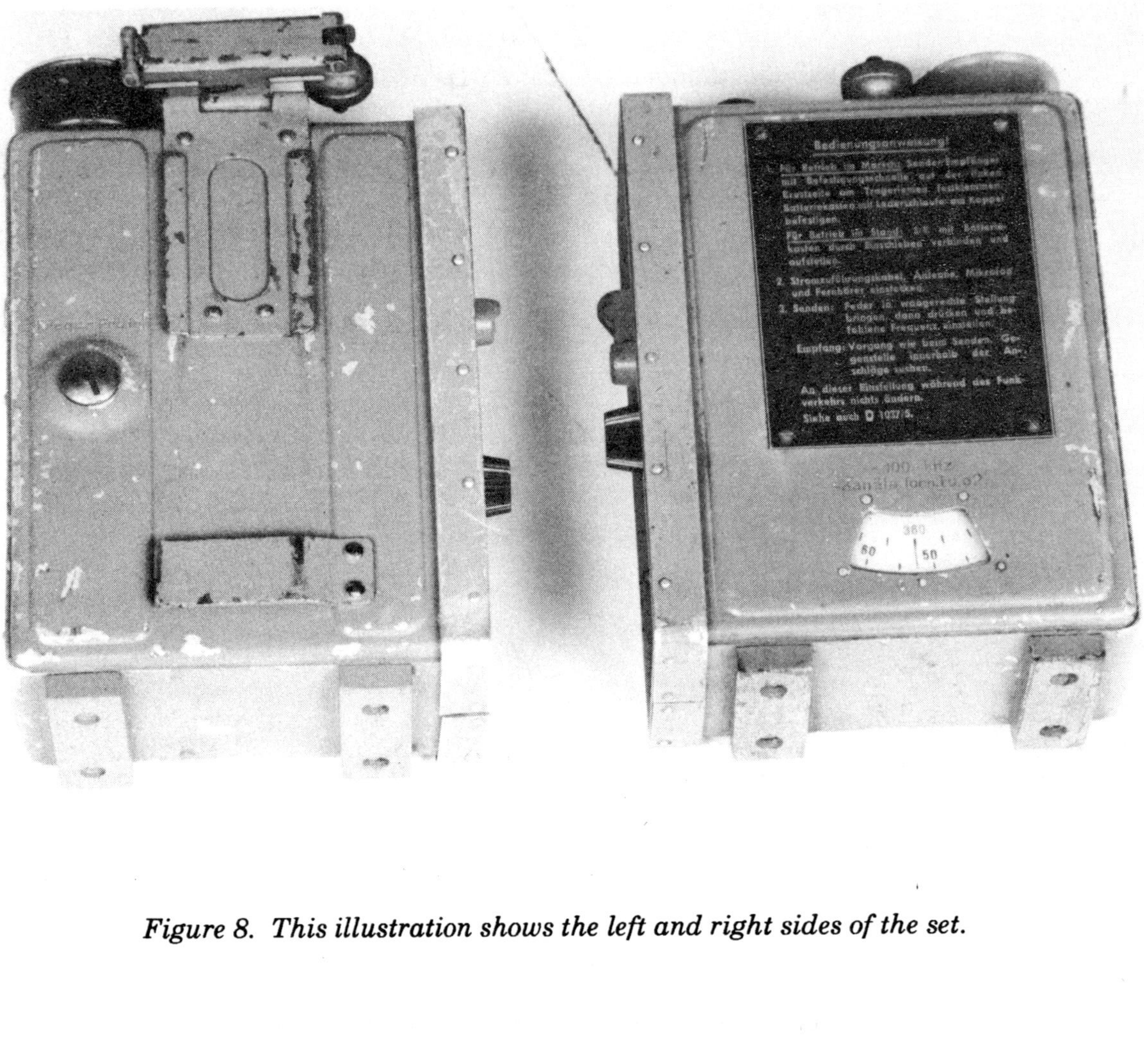

Figure 8. This illustration shows the left and right sides of the set.

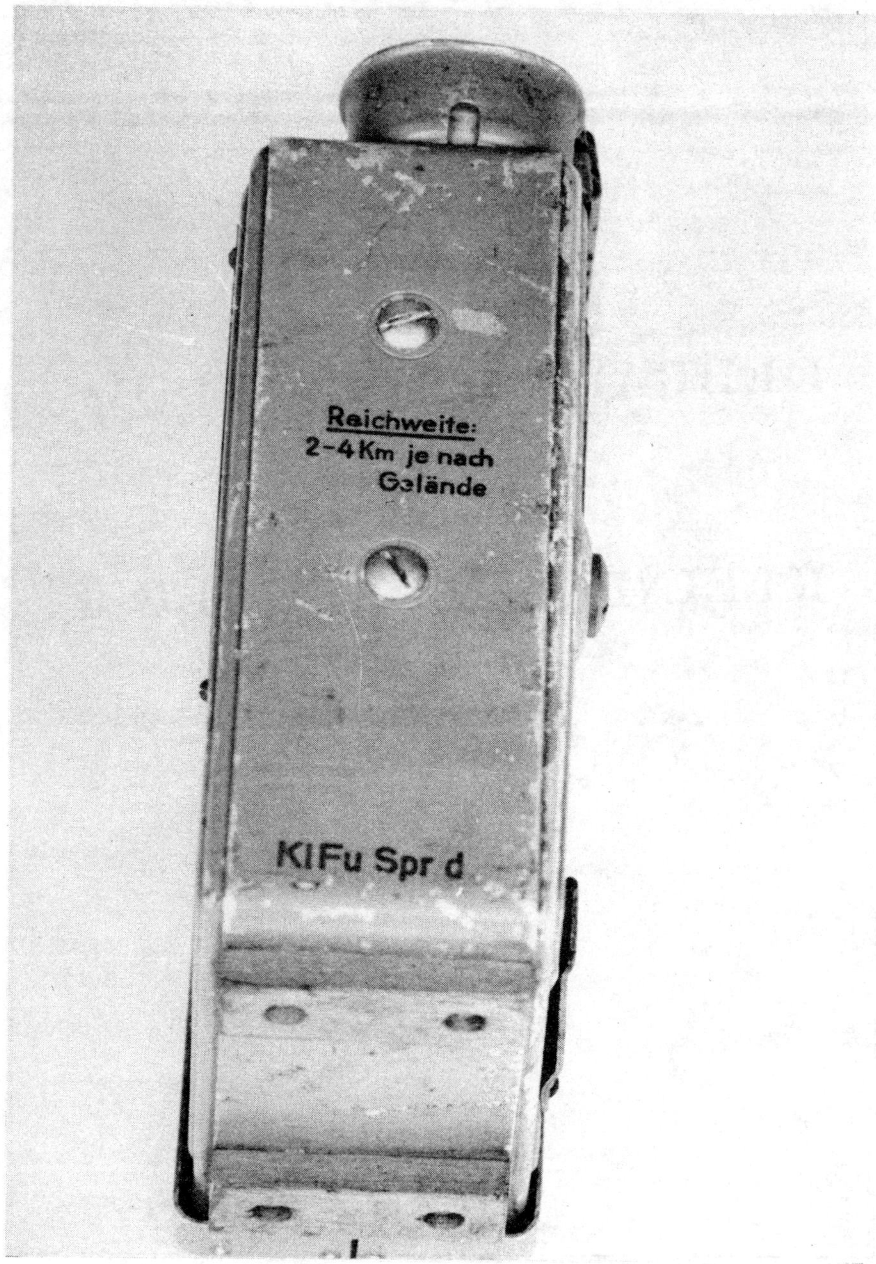

Figure 9. Rear view of the case.

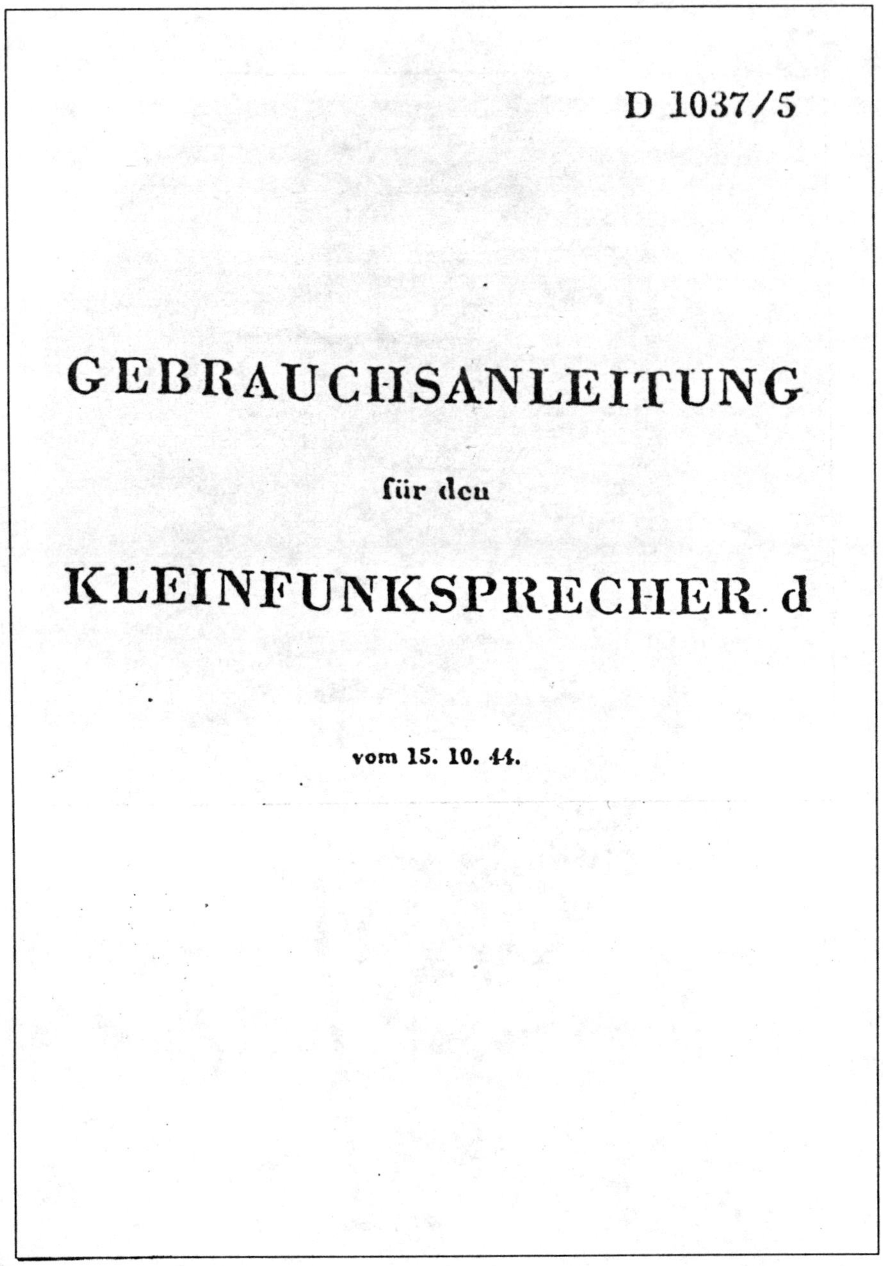

D 1037/5

GEBRAUCHSANLEITUNG

für den

KLEINFUNKSPRECHER d

vom 15. 10. 44.

Figure 10. Original manual for the Kleinfunksprecher d. These radios were issued in limited number in late 1944 and are one of the rarest of radio systems to be found.

VII

The 100 W.S. Transmitter

The 100 W.S., with telegraph key and hand microphone.

THE 100 W.S. TRANSMITTER

1. DESCRIPTION: The German radio transmitter **100 W.S.** (100-watt transmitter)* is used for **Telefonie** (voice) and **Telegrafie** (CW) transmission. The transmitter is commonly used in conjunction with German radio receivers **Torn E. b.** (portable receiver b) or **Spez. 976 Bs** (special receiver 976 Bs). It can be operated in a moving vehicle or as a fixed station and keyed from a remote location if desired. The set is contained in an aluminum-reinforced wooden case provided with a carrying handle. The cover is held in place by four hinged clasps. To obtain efficient operation at low frequencies, the antenna used with the **100 W.S.** will necessarily be bulky and hence rather difficult to install. The power for operating the set can be supplied by a gasoline-engine-driven generator, a 12-volt storage battery through a dynamotor, or from a-c mains through suitable rectifier equipment.

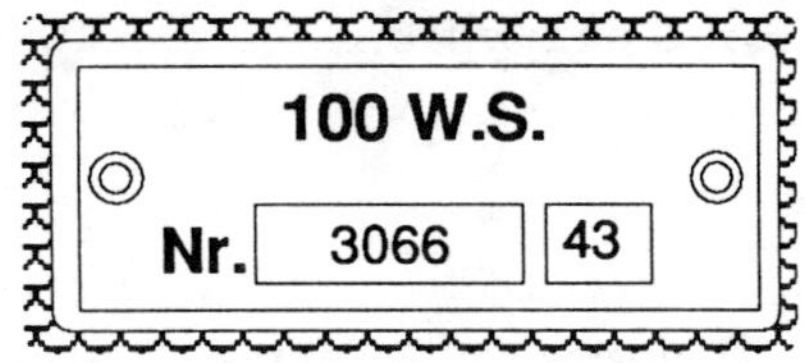

Figure 1. Nameplate of German radio set 100 W.S.

2. PERFORMANCE DATA: The table below lists the general characteristics and performance date of the 100 W.S. (100-watt transmitter).

PERFORMANCE DATA

Frequency range		200 to 1,200 Kc approximately.
This frequency range is broken up into 4 bands:		
red		200 to 310 kc, approximately
green		310 to 490 kc, approximately
blue		490 to 770 kc, approximately
yellow		770 to 1,200 kc, approximately
Dial graduation		kilocycles
Types of signals emitted		cw and voice
Type of modulation		amplitude
Method of modulation		grid
Type of transmitter		master oscillator-power amplifier (MOPA)
Number of tubes:		
Total		3
	Used for cw	2
	Used for Voice	3
	Master oscillator tube	1 type RS237
	Power amplifier tube	1 type RS237
	Modulator tube	1 type RS241
Power Output:		
High power:	Cw	100 watts
	Voice	25 watts
Low power:	Cw	10 watts
	Voice	no low-power position
Power requirements:		
	Plate	1,000 v, 240 to 300 ma, approximately
	Filament	12 v, 7.3 amp

Power supply:

Vehicular	12-volt storage battery through dynamotor unit U.100a. Current drain 33 to 40 amp
Stationary	gasoline-engine-driven generator type C

Antenna types .... 33-foot vertical, telescopic, winch with umbrella top; 20-foot sectional mast with umbrella top; roof type; 18-foot rod; or long wire

Distance range (approximate):

Type of antenna	Cw	Voice
33-foot winch mast	125 miles	45 miles
20-foot sectional mast	100 miles	30 miles
Roof or 18-foot rod antenna	25 to 50 miles	6 to 12 miles

3. ANTENNAS:

a. **General.** The frequency range in which the **100 W.S.** operates is very low. An antenna of comparatively large dimensions must therefore be used if maximum radiation efficiency is to be realized. Vehicular antennas are too short for maximum efficiency at frequencies of 200 to 1,200 kilocycles. For this reason, vehicular operation will not provide as much distance range as fixed operation where it is possible to erect an antenna large enough for the frequency employed. A few types of antennas for use with the **100 W.S.** are discussed in the following paragraphs.

(1) STATIONARY OPERATION. For stationary operation, a 33-foot telescopic winch antenna mast is a basic component of the antenna system. Because of its massive construction, this mast should be securely guyed when erected (fig. 2). The antennas given below are satisfactory for stationary operation with the **100 W.S.**

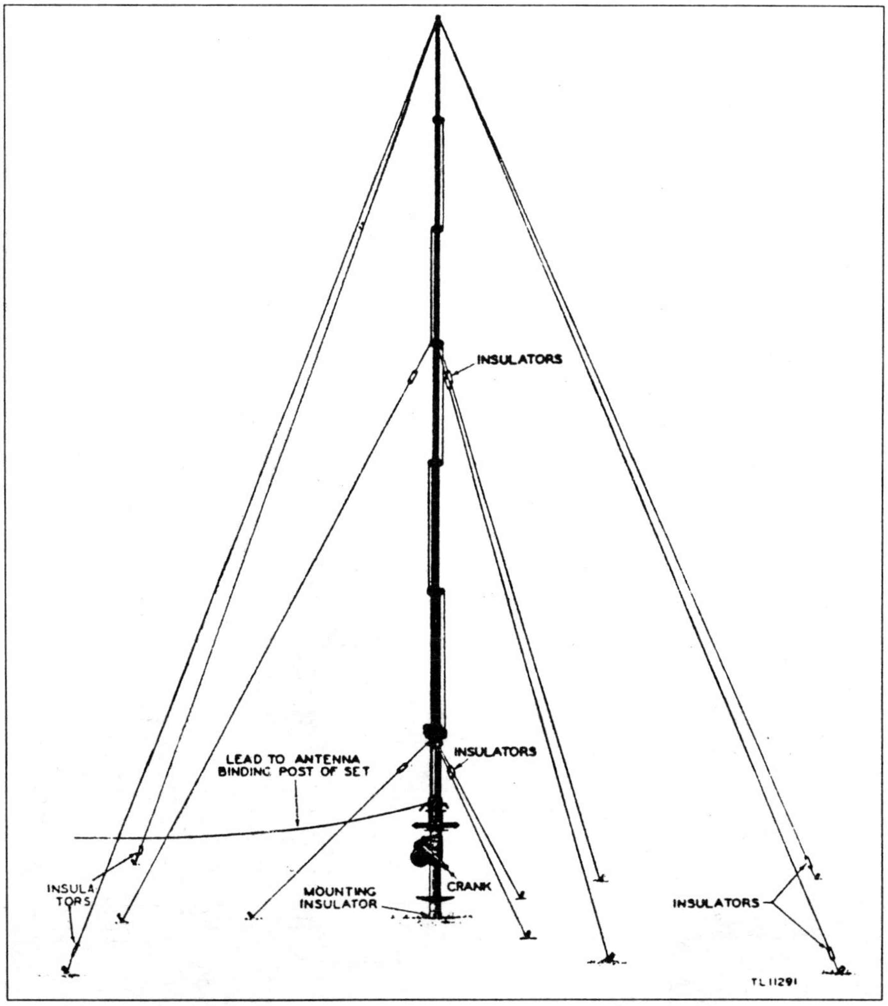

Figure 2. The 33 foot winch antenna mast installed.

(a) A 33-foot winch mast with a four-spoke umbrella.

(b) A 20-foot sectional mast with a three-spoke umbrella.

(c) A 33-foot winch mast with four antenna wires approximately 50 feet long.

(d) A 20-foot sectional mast with four antenna wires approximately 50 feet long.

(e) An open wire antenna 50 to 200 feet long.

(f) A rod antenna about 18 feet long.

(2) VEHICULAR OPERATION. When the **100 W.S.** is used in a moving vehicle, a mast section of 18 feet is necessary. Because of the limited length of antenna which can be used in vehicular installations, operation in higher frequency ranges of the **100 W.S.** is recommended.

(3) COUNTERPOISE. When the set is operating in a stationary position, the counterpoise for the above antennas should consist of four cables, approximately 82 feet in length, arranged radially around the base of the antenna mast. When the set is used for vehicular operation, the Gegengewicht (counterpoise) binding post should be connected to the chassis of the vehicle.

4. POWER SUPPLY:

a. **Stationary Operation.** German engine set type C a gasoline-engine-driven generator which delivers 1,000 volts for plate supply and 12 volts for filament supply is used to power the set. The German engine set type A may be used as a substitute, in which case the distance range will be reduced about 25 percent because of the lower plate voltage delivered (800 volts).

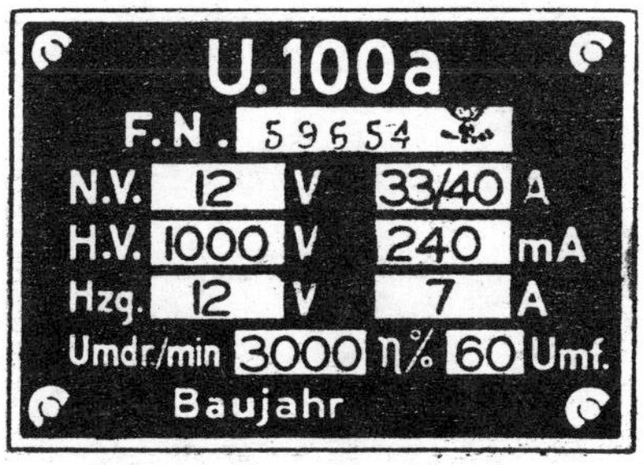

Figure 3. Nameplate of the U.100a Dynamotor.

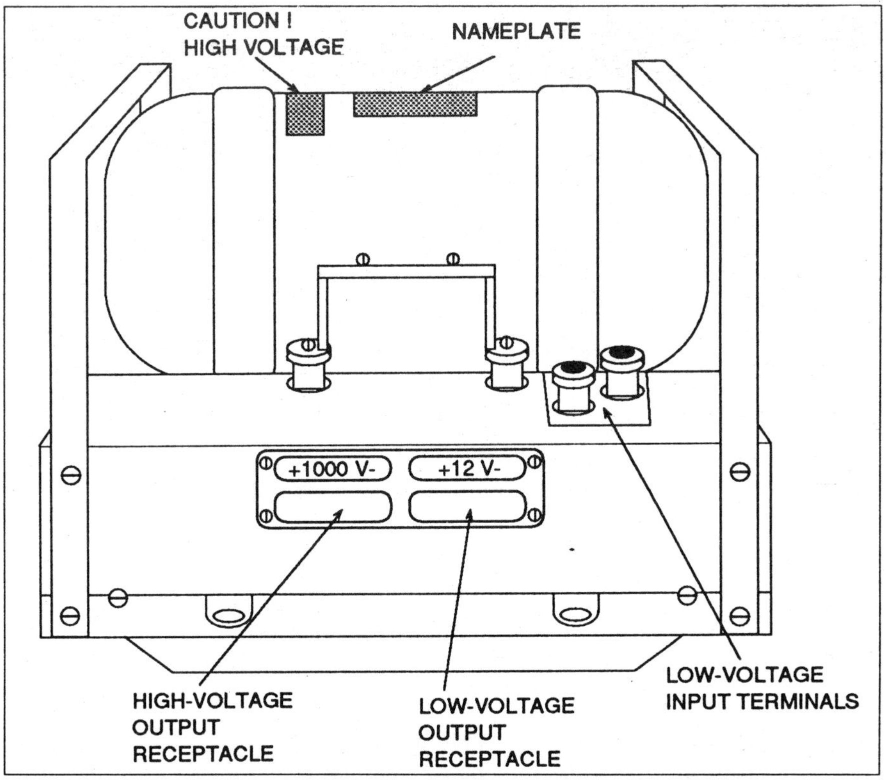

Figure 4. U.100a Dynamotor.

CAUTION:
The use of this equipment involves the use of high voltage (1,000 volts) which is dangerous to life. Observe at all times all safety regulations. Disconnect the power cables before removing the transmitter from the case.

b. Vehicular Operation. (1) Dynamotor unit **U.100** or **U.100a** may be used to supply the plate voltage necessary to operate the transmitter. Figure 3 shows the nameplate of dynamotor U.100a; figure 4 the dynamotor **U.100a**. The dynamotor operates from the 12-volt vehicular battery which also supplies the filament voltage for the transmitter. The German dynamotor **U.80a** may be used instead of the **U.100** or **U.100a** but the distance range will be reduced about 25 percent because of the lower plate voltage delivered (800 volts).

5. INSTALLATION:

a. Location.

(1) When installing equipment for operation, locate the units so that cable connections can be made conveniently. Bear in mind that the length of all cables is to be kept at a minimum. The antenna cable, in particular, is to be as short as possible.

(2) When installing for stationary operation, locate the equipment so that it can be operated from a remote point.

(3) When making a vehicular installation, be sure that the transmitter and dynamotor are securely mounted.

b. Cording for Stationary Installation.

(1) Before making any connections, set the following switches and controls as directed.

(a) Set the main switch on the transmitter to **Aus** (off) position.

(b) Turn the control marked **Heizung** (filament rheostat) to the extreme left.

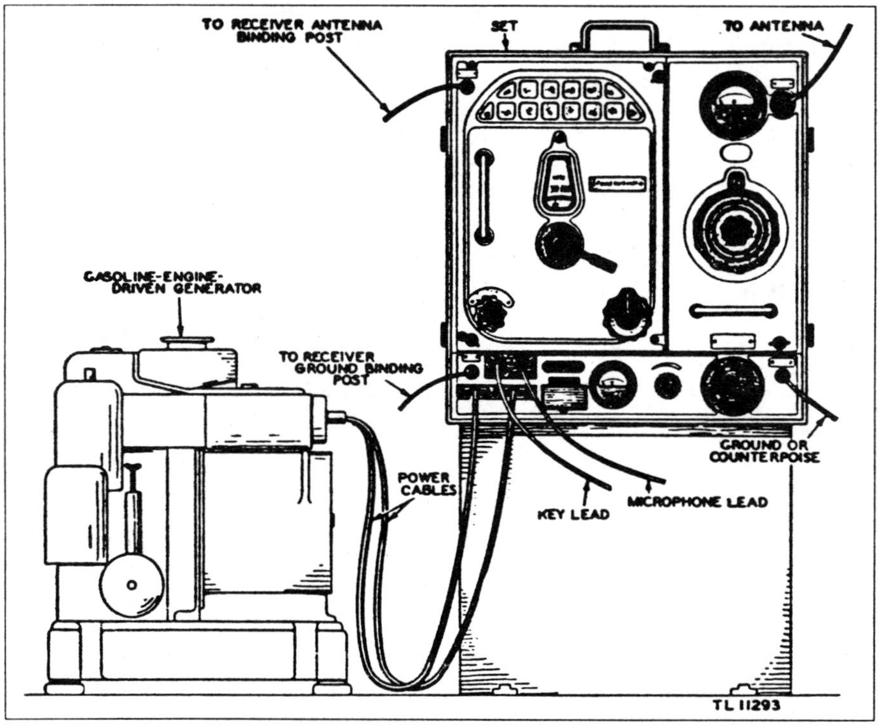

Figure 5. 100 W.S., cabled for stationary operation.

(c) Set switch marked **Leistung** (power) to **Klein** (low) position.

(2) To install gasoline-engine-driven generator type **C** or type **A** (fig. 5), proceed as follows:

(a) Connect the filament power cable to the terminals marked **+12V—** on the lower left of the transmitter panel and to the low-voltage terminals on engine set type **C**, or engine set type **A**.

(b) Connect the high-voltage power cable to the terminals marked **+ 1000V—** on the transmitter and to the high-voltage terminals on the power supply.

(3) To install Power Unit PE—49—(&), proceed as follows:

(a) Obtain negative low voltage through suitable plug at contacts numbered 53, 47 and 56, and positive low voltage at contacts numbered 45, 46, and 54 of the multiple-connector output receptacle on the power unit (fig. 6). Connect this low-voltage supply to the terminals marked **+12V—** on the lower left of the transmitter panel.

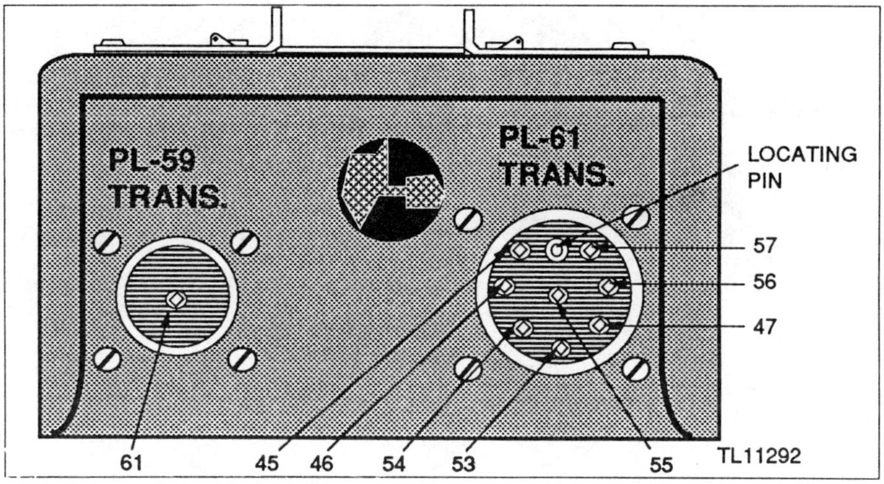

Figure 6. Output receptacles on Power Unit PE—49-(&) and Dynamotor BD-77-(&).

(b) Obtain negative high voltage at contact 55 of the multiple-connector receptacle, and positive high voltage, through a suitable plug, at contact 61 of the single-pin output receptacle on the power unit. Connect the high-voltage supply to the terminals marked **+1000V—** on the lower left of the transmitter panel.

(4) Connect the antenna to the terminal marked **Antenne** (antenna). Connect a wire between the terminal marked **Gegengewicht** (counterpoise) and a ground rod. Because of the high voltage present, a good ground connection is necessary.

(5) Connect the receiver antenna lead to the terminal on the upper left of the transmitter labeled **Empfanger Antenne** (receiver antenna). Connect the receiver counterpoise to the terminal labeled **Empfanger Erde** (receiver ground).

(6) Plug the microphone into the jack marked **Mikr.** (microphone). See figure 7.

(7) Plug the telegraph key into the jack marked **Taste Ort** (local key). Any telegraph key fitted with a suitable plug may be used.

c. **Cording for Vehicular Installation.**

(1) Before making any connections, set the following switches and controls as directed below. (Refer to figures 1, 6, 8, 9, 10)

(a) Set the main switch on the transmitter to **Aus** (off) position.

(b) Turn the control marked **Heizung** (filament rheostat) to the extreme left.

(c) Set the switch marked **Leistung** (power) to **klein** (low) position.

(2) To install dynamotor **U.100** or **U. 100a**, proceed as follows:

(a) Connect the 12-volt input terminals of the dynamotor to the 12-volt vehicular battery. Make sure that the battery contacts are clean and that the cable connects directly to the battery terminals.

Figure 7. From left to right, emphanger erde (receiver binding post), upper panel, Taste (key jacks) and Mikrofon, (microphone). Below is the power input receptacle, power switch, and Tastrelais (key relay) and voltmeter.

(b) Connect the filament power cable to the terminals marked **+12V—** on lower left of transmitter panel and to the low-voltage output receptacle of the dynamotor **U.100a** (fig. 4). On dynamotor **U.100** the connection at the dynamotor is to the 12-volt supply terminals.

(c) Connect the high-voltage power cable to the terminals marked **+1000V—** on the transmitter and the dynamotor.

(3) To install Dynamotor BD—77—(&), proceed as follows:

(a) Obtain negative low voltage through a suitable plug at contacts numbered 53, 47, and 56, and positive low voltage at contacts numbered 45, 46, and 54 of the multiple-connector output receptacle of the dynamotor unit . Connect this low-voltage supply to the terminals marked +12V— on lower left of the transmitter panel.

(b) Obtain negative high voltage at contact 55 of the multiple-connector receptacle, and positive high voltage through a suitable plug at the single-pin output receptacle of the dynamotor. Connect high-voltage supply to the terminals marked **+1000V—** on lower left of the transmitter panel.

(c) To operate the starting relay connect a switch as follows: through leads of the required length, connect one terminal of the single-pole single-throw switch to contact 57, and the other terminal of the switch to contacts 53, 47, and 56 of the plug used to make connection with the multiple-connector output receptacle.

Figure 8. These illustrations shows the data plate, filament rheostat, main switch, and Gagengewicht (counterpoise binding post).

(4) For vehicular installation, connect the roof antenna to the terminal marked **Antenne** (antenna), and a wire from the terminal marked **Gegengewicht** (counterpoise) to the chassis of the vehicle. For stationary operation connect the **Gegengewicht** (counterpoise) terminal to a ground rod. Because high voltage is present, be sure that a good ground connection is made.

(5) Connect the receiver antenna lead to the terminal marked **Empfanger Antenne** (receiver antenna), and the receiver ground binding post to the terminal marked **Empfanger Erde** (receiver ground).

6. a. Voice Operation

(1) Set the main switch on the transmitter to the **Telefonie** (voice) position.

(2) Turn the control marked **Heizung** (filament rheostat) to the right, (fig. 8).

(3) Press the microphone switch and speak into the microphone, slight upward deflections on the antenna meter indicate modulation.

(4) Loosen the retaining screw on the dial lock and select the desired band. This is done by first obtaining the frequency setting by means of the color coded band switch. Frequency range is from 200 to 1,200 kc. For fine tuning of the desired frequency, turn the tuning control knob located near the lower right side of the access door. Dial in the desired frequency (fig. 9).

(5) The power switch is not effective when the main switch is in the **telephonie** (voice) position.

CAUTION: Do not attempt to alter setting of the band switch and the switch marked Antennen - Stufenschalter (Antenna Tapping Switch) when the main switch is at the Telephonie (voice) position. This may cause arcing at the switch contacts. To return to transmitter, turn the main switch to telegrafie. (cw).

Figure 9. From top to bottom, dial, band switch, dial lock (lower left), right tuning control (lower right).

(6) Press the blue button labeled 1500V on the voltmeter and check the plate voltage. The reading should be between 800 and 1,000 volts, indicated in the blue section of the seals.

(7) Turn the control marked **Festellung** (antenna turning lock) to release the antenna control.

(8) Depress the key and turn the control marked **Antennen-Abstimmung** (antenna tuning) until maximum deflection is obtained on the antenna meter **A**. Ammeter deflection may be improved sometimes by turning the switch labeled antennen-**Stufenschalter** (antenna tapping switch) until a different figure of the same color appears in its window. When a satisfactory setting is found, turn the control marked **Festellung** (antenna tuning lock) to the extreme right, holding the control marked **Antennen-Abstimmung** (antenna tuning), to prevent disturbing the proper setting, see figure 10.

(9) Change over to receiver. Set the main switch to **Empfang** (receiver).

b. C-w Operation.

(1) Tune the transmitter as described in paragraph 6a. With the main switch set at the position marked **Telegrafie** (cw), use the key for c-w transmission.

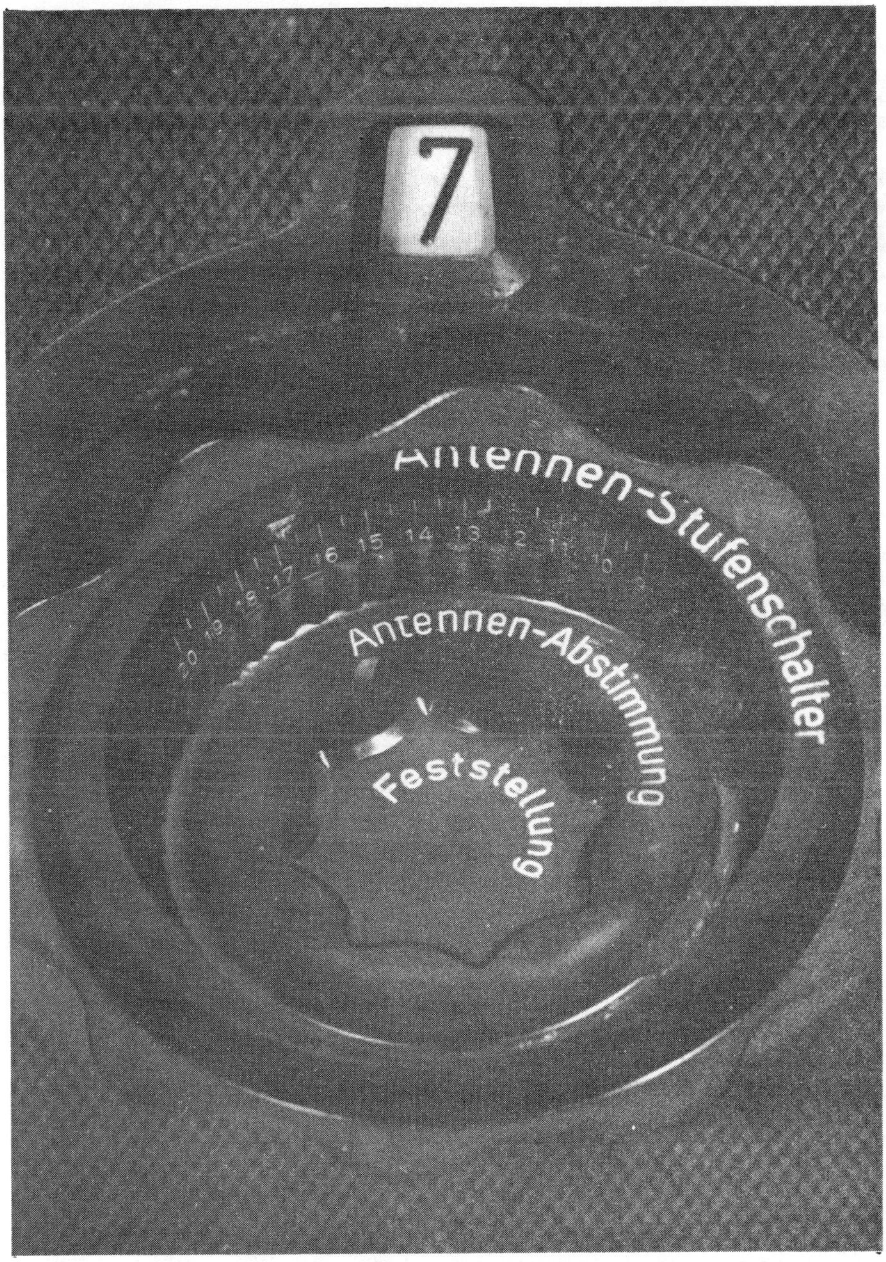

Figure 10. Close up view of antenna control dials.

CAUTION: When power switch has been changed over to gross (high) position, do not attempt to alter the settings of the band switch and the switch marked Antennen-Stufenschalter (antenna tapping switch). This may cause arcing at the switch contacts.

(2) To operate at maximum power output, turn the switch marked **Leistung** (power) to **Gross** (high) position.

(3) To operate the transmitter from a remote location, plug the remote key line into the jack marked **Taste Fern** (remote keying). Refer to paragraph 7 for details.

7. REMOTE OPERATION:

a. **General.** For remote operation of the **100 W.S.**, follow installation and operating instructions described in paragraphs 5 and 6. Provide the additional equipment needed to key or modulate the transmitter shown in figures 11, 12, or 13. The contacts of the remote key jack marked **Fern** (remote) are connected to the filament voltage supply through the coil of the remote key relay. The key relay is located under the cover marked **Tastrelais** (key relay). A telegraph key used in the remote key jack operates the key relay. The key-relay contacts key the transmitter in the same circuit as a key used in the jack marked **Ort** (local). The 22 1/2-volt external battery is used in series with the field wire line to provide ample current to operate the key relay when more than 1/4 mile of field wire line is used. The receiver used during remote operation should be at the remote point.

b. Remote Keying.

(1) Use a voltmeter to determine the polarity of the contacts of the remote key jack marked **Fern** (remote). Figure 11 shows the polarity of the set 100 W.S.

(2) Connect the negative contact of the remote key jack to the positive side of the external battery. Connect one of the field wire lines to the negative side of the external battery, and the other to the positive contact of the remote key jack.

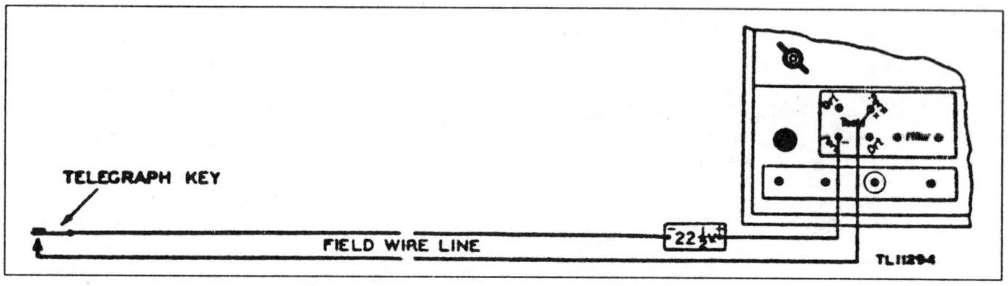

Figure 11. Circuit providing remote keying of the 100 W.S.

c. Remote Keying and Communication Between Operators.

(1) Connect the external battery, the remote key, two Telephones EE—8—A, and two 1.0 µf (or larger) capacitors as shown in figure 12. This arrangement provides a means for remote keying of the transmitter and for field telephone communication between the operator and the remote operator.

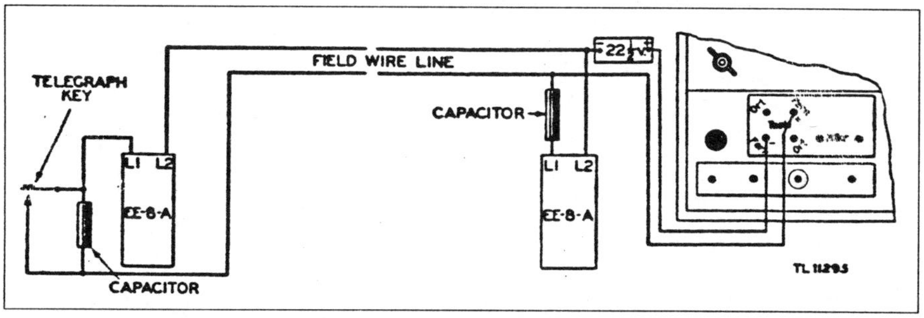

Figure 12. Circuit providing remote keying of the 100 W.S. and communication between operators.

d. Modulation Using Telephones EE—8—A.

(1) Turn the main switch to **Aus** (off) position. Remove the transmitter form the case (par. 10c) and remove screw-clamp connection No. 68 from wire-wound resistor No. 65. (This resistor is mounted near the center of the chassis, on the corner of the chassis partition behind the panel voltmeter.) Part numbers are enclosed by a circle or a rectangle. The No. 68 referred to above is stamped on the screw-clamp connection. After this connection is removed from the resistor, tape it to prevent a short circuit and replace the transmitter in the case.

(2) Connect the remote line to the jack marked **Mikr.** (microphone) through a 1.0 µf (or larger) capacitor as shown in figure 13.

(3) The connection No. 68 removed in subparagraph d (1) above must be replaced for most efficient operation when the regular microphone is used for modulating the transmitter.

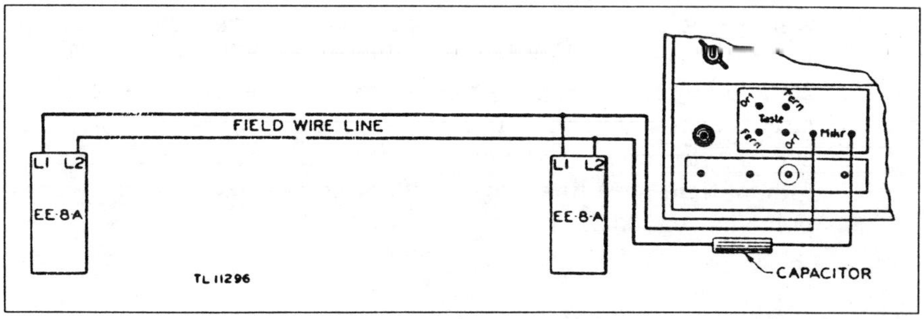

Figure 13. Connection of Telephones EE-8-A to modulate the 100 W.S.

8. PICTURE TRANSMISSION: On the lower left of the front panel above the jack marked **Mikr.** (microphone), is a jack used for connecting German facsimile equipment to the transmitter. If such equipment is available, and to be used, remember signal security would be impossible. A study of available equipment will suggest the possibilities of its use with this transmitter.

9. CALIBRATION: The frequency range of the **100 W.S.** is covered in four bands which are switched in by the band switch whose color code is the same as the color code which appears in the tuning dial window. The tuning dial is calibrated in kilocycles. The set is so constructed that its frequency will be very stable. If the calibration is off, however, do not attempt to recalibrate; merely set the **100 W.S.** on the frequency desired by using a frequency meter, or by zero-beating the **100 W.S.** to a receiver which is set on the correct frequency.

10. MAINTENANCE: Complete maintenance instructions are not included in this bulletin. The following information, however, should prove helpful in locating trouble when the set fails to operate.

a. Check the filament and plate voltages with the panel voltmeter. Check the input supply voltage with the transmitter turned on.

b. Check all cords, plugs, and connections.

c. If the antenna ammeter shows no deflection, it may indicate a defective power-amplifier or master-oscillator tube. Replace these tubes in the order names above, checking for operation after each tube is replaced. To change the tubes, it is necessary to remove the transmitter from its case. Do this by loosening the three retaining screws encircled in red on the front panel of the transmitter, and pulling the transmitter forward out of the case. Figure 14 is a top rear view of the transmitter showing the position and function of the tubes. Lift the retaining ring and turn it away from the tube to be removed. Remove either the power amplifier or the master oscillator tube by turning it to the left until it is loose; then lift it from the socket. To remove the modulator tube, lift it vertically from the socket.

Figure 14. The 100 W.S. Top rear view removed from case.

Figure 15. This illustration shows the cover in place; the holes on the top of the case are for air circulation.

d. If c-w operation is normal and speaking into the microphone does not cause change in deflection of the antenna ammeter, or if the distant station reports poor quality speech, check or replace the microphone and microphone cable, and the modulator tube.

e. If the distant station reports poor quality cw, check the key cable and clean the key contacts.

f. If tuning is normal but the antenna ammeter does not return to zero during keying when the key is released, the fault may be in the key, the key cable, or the remote key relay. If either of these components is defective, the antenna ammeter will return to zero when the defective component is disconnected from the transmitter. To remove the remote key relay, which is a plug-in type, loosen the screw on the bottom side of the cover marked Tastrelais (key relay), raise the cover, and pull the relay from its receptacle. Only competent personnel should attempt adjustment or repair of this relay.

g. If operation while using the local key is satisfactory and remote key operation impossible, connect the local key to the remote key jack. If operation is possible, the fault is in the remote key, remote key line, or the external batteries. If operation is still impossible, the remote key relay is defective.

Figure 16. The 100 W.S., front view.

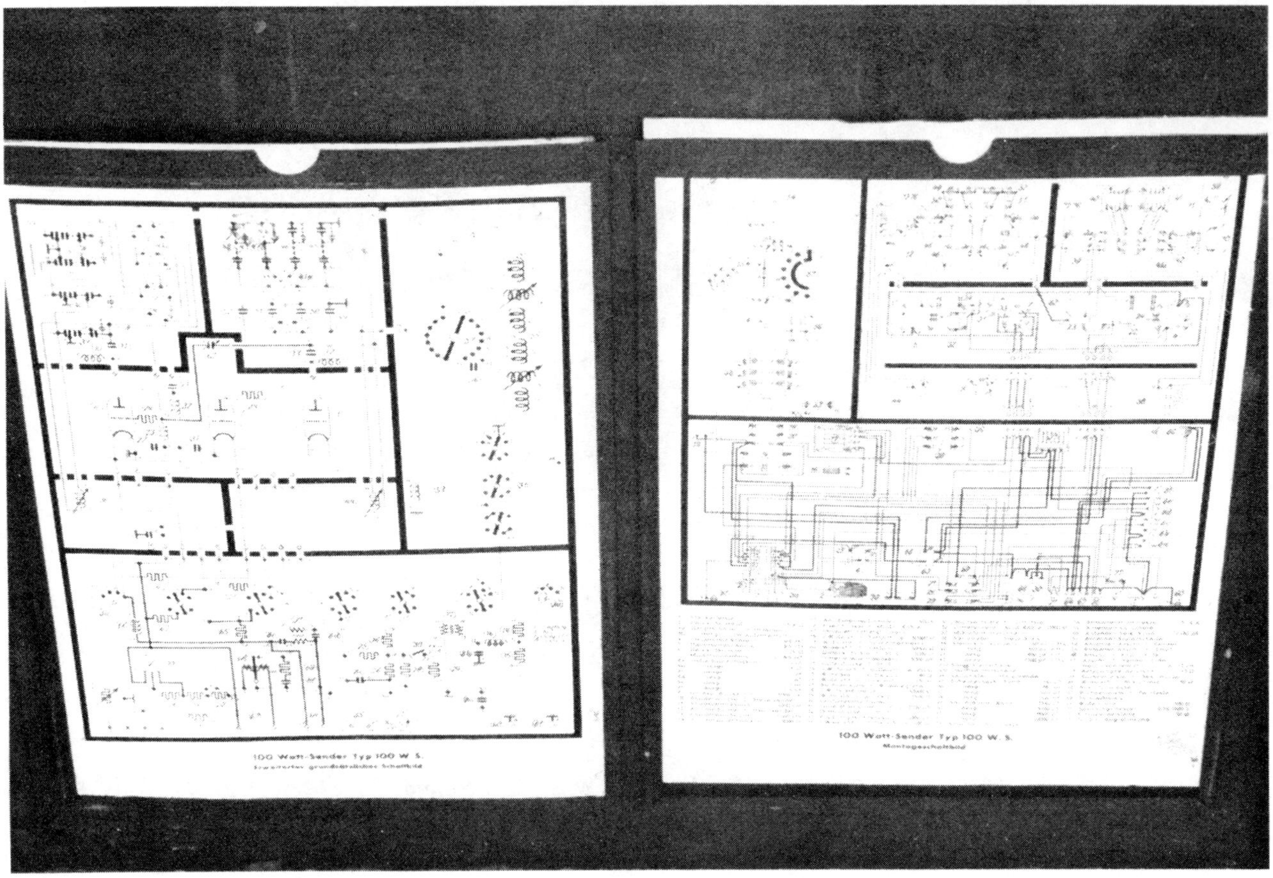

Figure 17. Close-up view of the wiring schematics on the inside of the cover.

Figure 18. Command and control vehicles such as this Opel-Blitz A-Type radio telephone truck were configured with the more powerful radio sets such as the 100 Watt Sender and the short wave Receiver Kw.E.a. Several other types of wheeled vehicles were used throughout the war for communications equipment.

Figure 19. Rear view of a kfz 17 Funkkraftwagon (wireless car). Note the 33 ft. winch antenna mast and six hand held cable dispensers.

VIII

The Torn. Emphanger b - Receiver

Figure 1. A signal team gathers around a Torn. E.b during the battle for France.

The Torn. Emphanger b - Receiver

1. DESCRIPTION: The German radio receiver Tornister Empfanger b, abbreviated Torn. E. b.(b), is designed for direction-finding installation, to be set up on the ground or in a vehicle as an intercept receiver. This set is a pack-type, tuned radio-frequency, four-tube receiver, capable of receiving continuous-wave (c-w) and modulated signals over a frequency range of approximately 97 to 7,095 kilocycles. This set can be operated from a battery power supply contained in the accessories case or Umformersatz EWc (vibrator power supply type EWc). The vibrator power supply operates from a 12-volt storage battery. The apparatus and accessories cases are water-tight and can bolted firmly together for transportation by one person. Handles are provided for each section so that they can be carried separately if desired.

2. PERFORMANCE DATA: The table below lists the performance and general characteristics of the Torn. E. b. (portable receiver b).

PERFORMANCE DATA

Frequency range:		97 to 7,095 kc approximately (covered in 8 overlapping bands.) This frequency range is broken up into 8 bands: 1. 96.6 - 177.8 kc 2. 171.0 - 313.7 kc 3. 304.0 - 558.8 kc 4. 540.0 - 990.0 kc 5. 955.0 - 1740 kc 6. 1674 - 3075 kc 7. 2920 - 4820 kc 8. 4360 - 7095 kc
Types of signals which may be received:		cw, tone, and voice
Type of receiver:		Tuned r-f with regenerative detector
Number of tubes:		4
	2	r-f amplifiers type RV2P800
	1	detector type RV2P800
	1	a-f amplifier type RV2P800
Sensitivity:		20 to 100 microvolts for 1 milliwatt output
Power supply:		Battery or vibrator
Battery:		
	Filament	2-volt, German type 2B 38 storage cell
	Plate	90-volt, German type DIN/VDE 1600 dry battery
Vibrator:		
	Filament	12-volt storage battery with dropping resistor built into battery pack
	Plate	90-volts delivered by vibrator supply
Current consumption:		
	Filament	approximately 800 milliamperes
	Plate	approximately 12 milliamperes
Antenna:		Long wire or any rod antenna

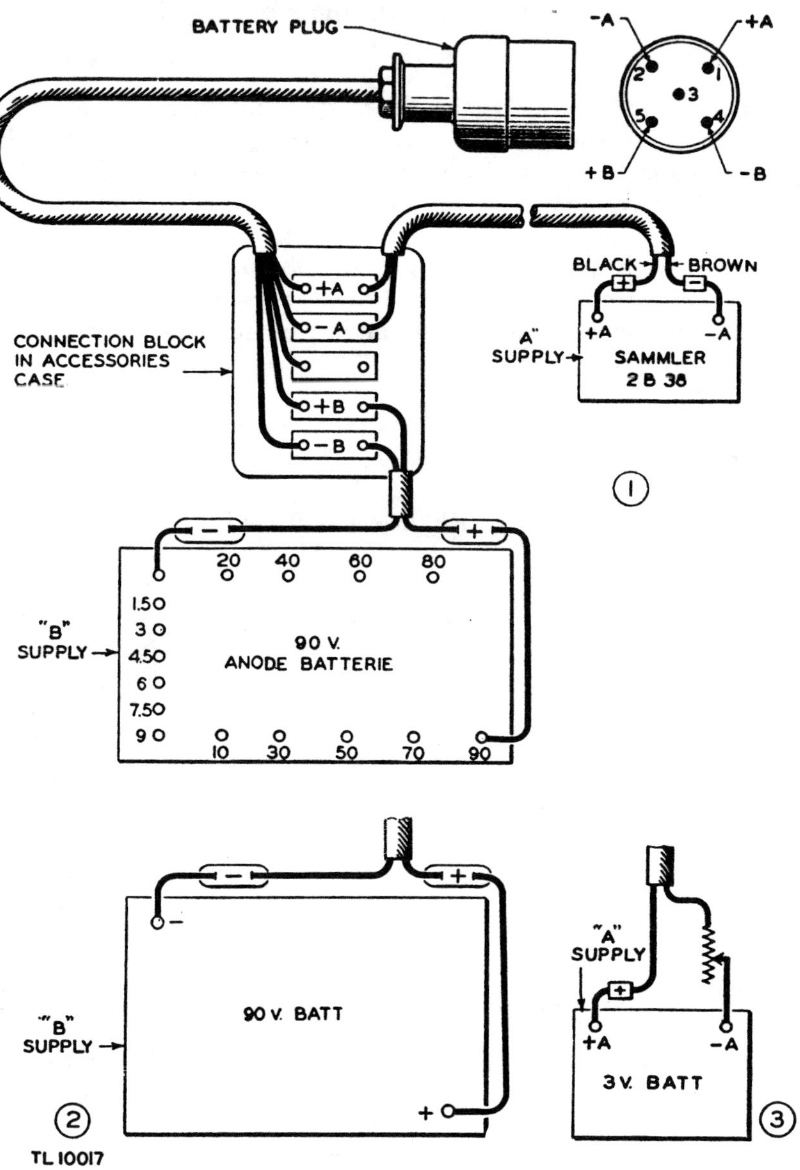

① German battery connections.
② Substitute American "B" supply.
③ Substitute American "A" supply.

Figure 2. Battery connections for the ***Torn. E.b.***

3. POWER SUPPLY INSTALLATION:

a. Battery Installation. The batteries should be connected in the accessories case as shown in figure 1. The storage cell, type 2B, will fit into the compartment on the right. **The Anode Batterie** (B battery) fits into a compartment on the lower side of the case (figure 2).

(1) FILAMENT SUPPLY SUBSTITUTES. When the German storage cell is unavailable, a 2-volt storage cell of any manufacture may be used. Dry cells of 3 volts may be used in place of the storage cell but a dropping resistor of approximately 1.25 ohms at 7 watts will have to be connected, as shown in Table 1.

Note: Adjust the value of the dropping resistor, installed when dry cells are used for filament supply, so that the voltmeter on the set reads 2. This will lengthen the life of the tubes by preventing the application of excessive voltage to the tube filaments. However, in some of the later models of the Torn. E. b. the voltmeter is omitted from the set. In these models an external meter must be used to assure that the correct voltage will be applied to the filaments.

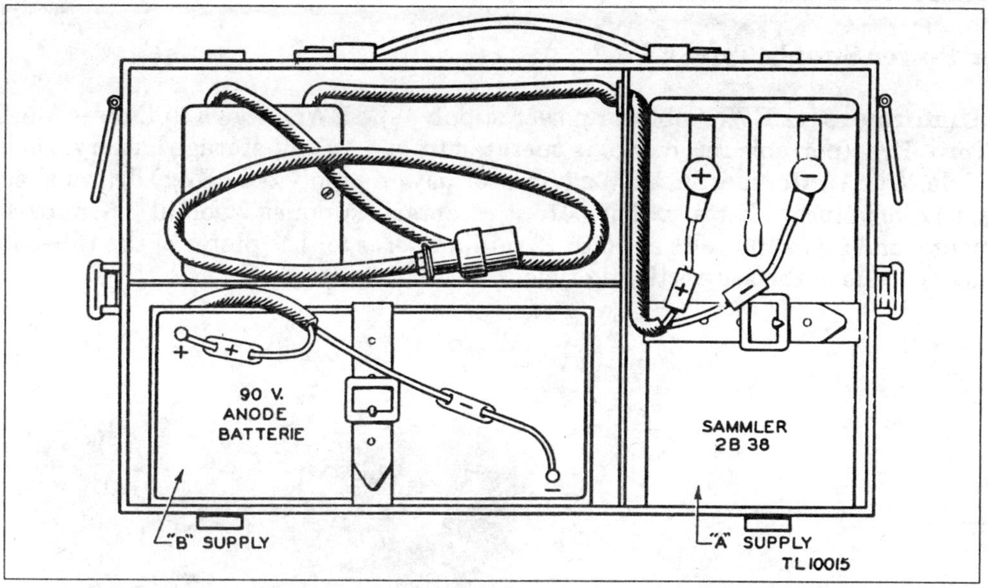

Figure X. Accessories case of German Radio set ***Torn. E. b.***

(2) PLATE SUPPLY SUBSTITUTES. When German **Anode Batteries** (B batteries) are not available, B batteries of American manufacture may be substituted, as shown in Table 2.

Battery type	Number used	Connection	Delivered voltage	Life
BA-23	4 6	Series-parallel Series-parallel	3 volts 3 volts	50 hours 110 hours
BA-35	4 6	Series-parallel Series-parallel	3 volts 3 volts	35 hours 65 hours
BA-65	2 4	Series Series-parallel	3 volts 3 volts	11 hours 35 hours
BA-15A	8 12	Series-parallel Series-parallel	3 volts 3 volts	35 hours 65 hours

Table 1

PLATE SUPPLY (B Battery)

Battery type	Number used	Connection	Delivered voltage	Life
BA-36	2	Series	90 volts	18.5 hours
BA-2	8	Series-parallel	90 volts	17 hours
BA-8	4	Series	90 volts	43 hours

Table 2

(3) BATTERY LIFE. The expected life of substitute American batteries is given in the table below. Other batteries which will supply 2 volts for the filament and 90 volts for the plate may be used. The batteries listed, however, will give reasonable life for a continuous filament drain of 800

milliamperes and a continuous plate drain of 12 milliamperes. Intermittent use of the set will, of course, greatly increase the life of the batteries. Refer to Tables 1 and 2 for expected life for the various substitute batteries.

b. Vibrator Power Supply Type EWc.

(1) The **Umformersatz EWc** (vibrator power supply type EWc) shown in figures 3-5, is used when the Torn. E. b. (portable receiver b) is operated from a 12-volt storage battery, such as is used in a vehicle. The **Umformersatz EWc** (vibrator power supply type EWc) draws about 1 ampere from a 12-volt storage battery. The output receptacle supplies 2 volt at 0.8 ampere for the filaments, and 90 to 100 volts at 10 to 12 milliamperes for the plates of the tubes in the set. The vibrator used is of the self-rectifying type.

Figure 3. Vibrator power supply E.W.c, for the Torn. E.b. Top view with the cover opened. Note the manufactures commercial label on the lid.

Figure 4. Another view of the vibrator power supply in the closed position.

(2) The 12 volts supplied is reduced to the proper value for the tube filaments by a series-dropping resistor contained in the vibrator power supply. The filament voltage is held to a steady value by an **Eisenwassers-toffwiderstande** (ballast tube) in series with the tube filaments and the dropping resistor. The filament circuit is protected by a **Sicherungen** 2 amp. (fuse, 2 amp).

(3) The cording for use of the vibrator supply is shown in figure 6.

c. Accessories Case.

(1) The container is divided into three compartments, as shown in figure 7. Components consist of a 2-volt lead storage cell, a 90-volt plate battery, two headsets, and the battery cable. The batteries are held firmly in place by leather straps. The storage cell is connected to the red (+) and blue (-) lugs of the battery cable. The two conductors from the cable to the plate battery are also labeled (-) and (+) and are connected to the (-) and (+) terminals of the 90-volt battery.

(2) In the front cover of the accessories case is an aperture which permits connection of the cable to the set without removing the cover. The lid which is provided for this opening is closed during transportation.

Figure 5. Side view of the E.W.c, showing the data plate.

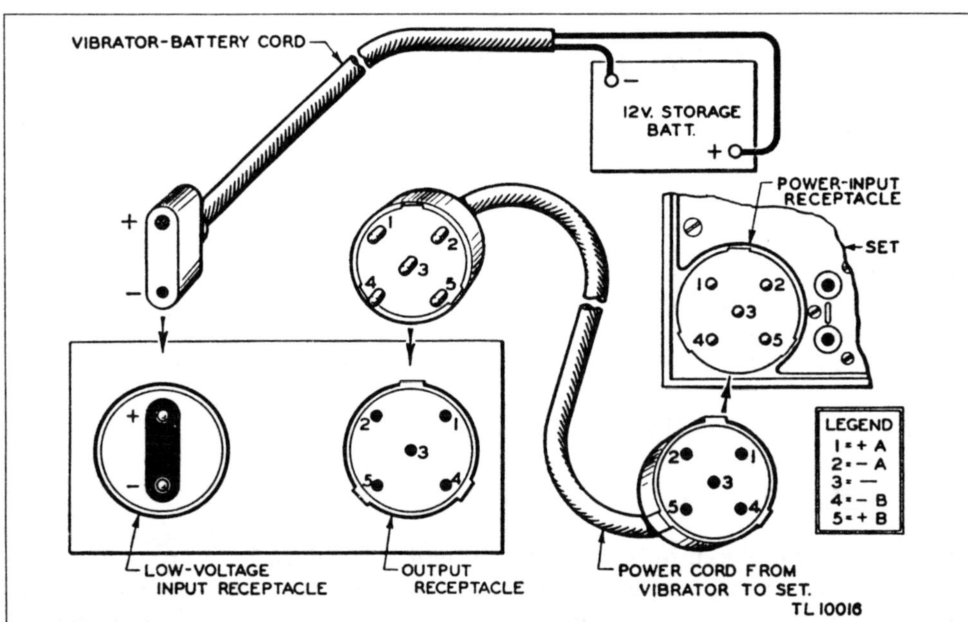

Figure 6. Cording diagram for use of the vibrator supply with the set.

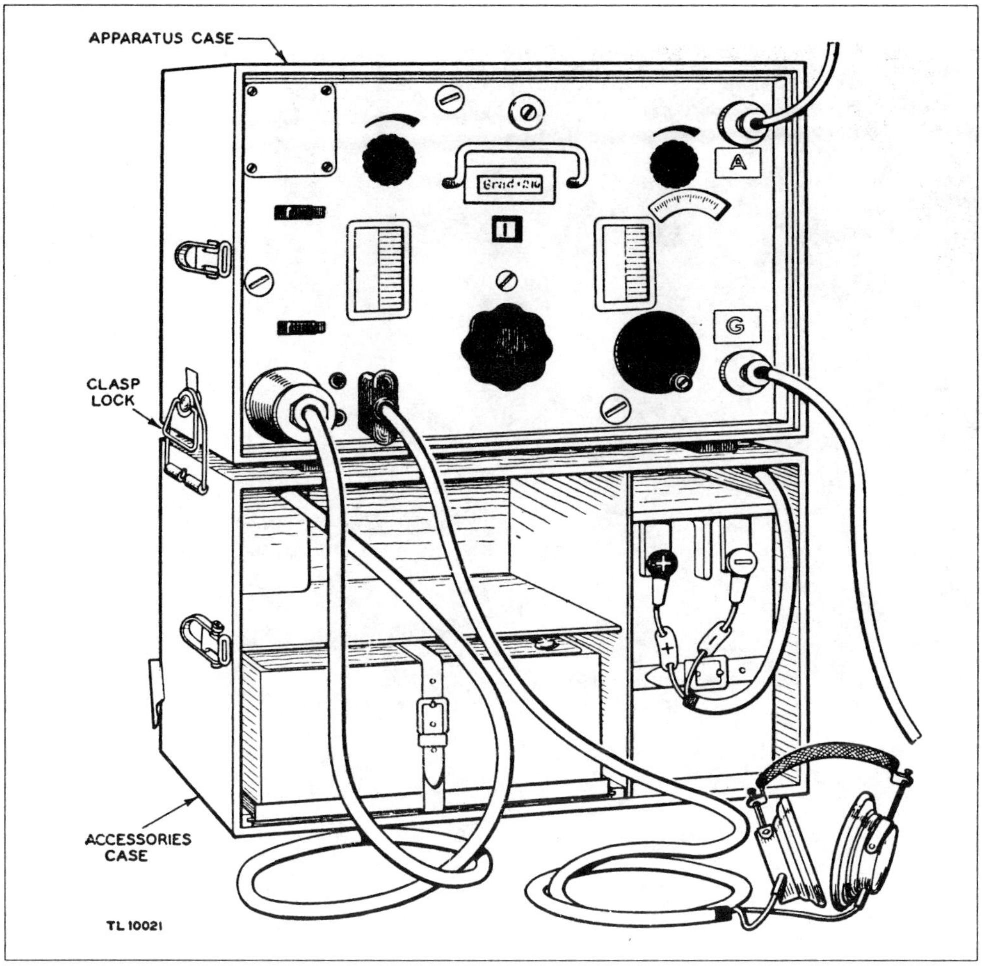

Figure 7. Line drawing showing the Torn E.B., set up for operation.

4. INSTALLATION:

a. Assembly. Place the apparatus case on top of the accessories case. These sections are secured together by clasp locks on the ends. To lock these sections in place see figures 7 and 8 and proceed as follows:

(1) Depress locking lever, grasp the triangular eccentric handle on the apparatus case, and rotate it until the small side of the eccentric is on top.

(2) Release the triangular eccentric handle on the apparatus case, grasp the loop on the accessories case, and place it over the eccentric on the apparatus case.

(3) Turn the triangular eccentric handle on the apparatus case a complete half-turn, so that the locking lever snaps into place.

(4) To unlock, depress locking lever and reverse procedure.

b. Antenna and Ground. Connect antenna to binding post marked **A** (antenna) and the ground or counterpoise to the binding post marked **G** (counterpoise), on the set (see figure 9). The antenna can be a 45-foot length of wire, one end of which may be attached high on a tree. Be sure to keep it insulated from the tree and free from contact with surrounding objects. Vertical rod antennas can also be used. Use a good ground; if one is not available, use a 45 to 65 foot counterpoise cable.

c. Ground Set. The following general instructions are to be used when the **Torn. E. b.** (portable receiver b) is operated from the batteries contained in the accessories case (see figure 10).

Figure 8. Side and bottom view of the case. The battery box is positioned beneath the receiver and held in place by the eccentric handles.

Figure 9. The right side of the set, close-up view from top to bottom, "A" antenna binding post, Rückkoppl regeneration control tuning dial calibration table, tuning control knob, and "G" counterpoise binding post.

Figure 10. Front view of the set, cover removed.

(1) Install and connect batteries as indicated in paragraph 3.

(2) Connect antenna and ground or counterpoise as indicated in paragraph 4b above.

(3) Insert battery plug into power input receptacle at lower left front of set.

(4) Plug the headset into any of the jacks marked **Fernhorer** (headphones). Align the plug with the white line. The set is now installed and ready for operation.

NOTE: The set may be installed as a ground set and may be operated from the vibrator power and a 12-volt storage battery. Some means for charging the battery should be available.

d. Vehicular Set. When using the **Torn. E. b.** as a vehicular set, be sure that the set is firmly secured to the vehicle. The **Umformersatz EWc** (vibrator power supply type EWc), which operates from a 12-volt storage battery, can be used to operate the set (see figure 11).

(1) Install and connect **Umformersatz EWc** to the storage battery and the set, as shown in figure 6.

(2) Connect vehicular antenna to **A** (antenna) jack on set.

(3) Connect a wire between vehicle chassis and **G** (counterpoise) jack of set.

(4) Plug the headset into any of the jacks marked **Fernhorer** (headphones) shown in figure 12. Align the plug with the white line. The set is now installed and ready for operation.

NOTE: The Torn. E. b. may be installed in a vehicle and operated from the batteries contained in the accessories case.

Figure 11. The Torn. E.b., and E.W.c., vibrator power supply.

5. OPERATION:

a. General. After the set is installed it is put into operation as follows (see figures 10, 13 and 14 for location of controls):

(1) Turn the **Aus-Ein** (off-on) switch on **Ein** (on).

(2) Test the battery voltages; by reading the voltmeter on the set the pointer should be in the red section of the scale for correct filament voltage. Upon pressing the button on voltmeter, the plate voltage is read and it should be in the blue section of the scale.

NOTE: On some of the later models of the Torn. E. b. this voltmeter is omitted. On these models check the batteries under load at their terminals with an external voltmeter when the receiver sensitivity is low.

(3) Turn **Frequenzeinstellung Grob** (band switch) to band 6, set the **Frequenzeinstellung Fein** (tuning control) between 80 and 100, and adjust **Anpass** (antenna trimmer) to maximum output. Either a received station or background noise may be used as an indication. This readjustment will only be necessary if the antenna is changed.

(4) Turn the **Lautst** (volume control) to maximum.

(5) Turn **Frequenzeinstellung Grob** (band switch) to desired frequency band.

(6) Set the control marked **Ruckkoppl** (regeneration) just below the point of oscillation for voice

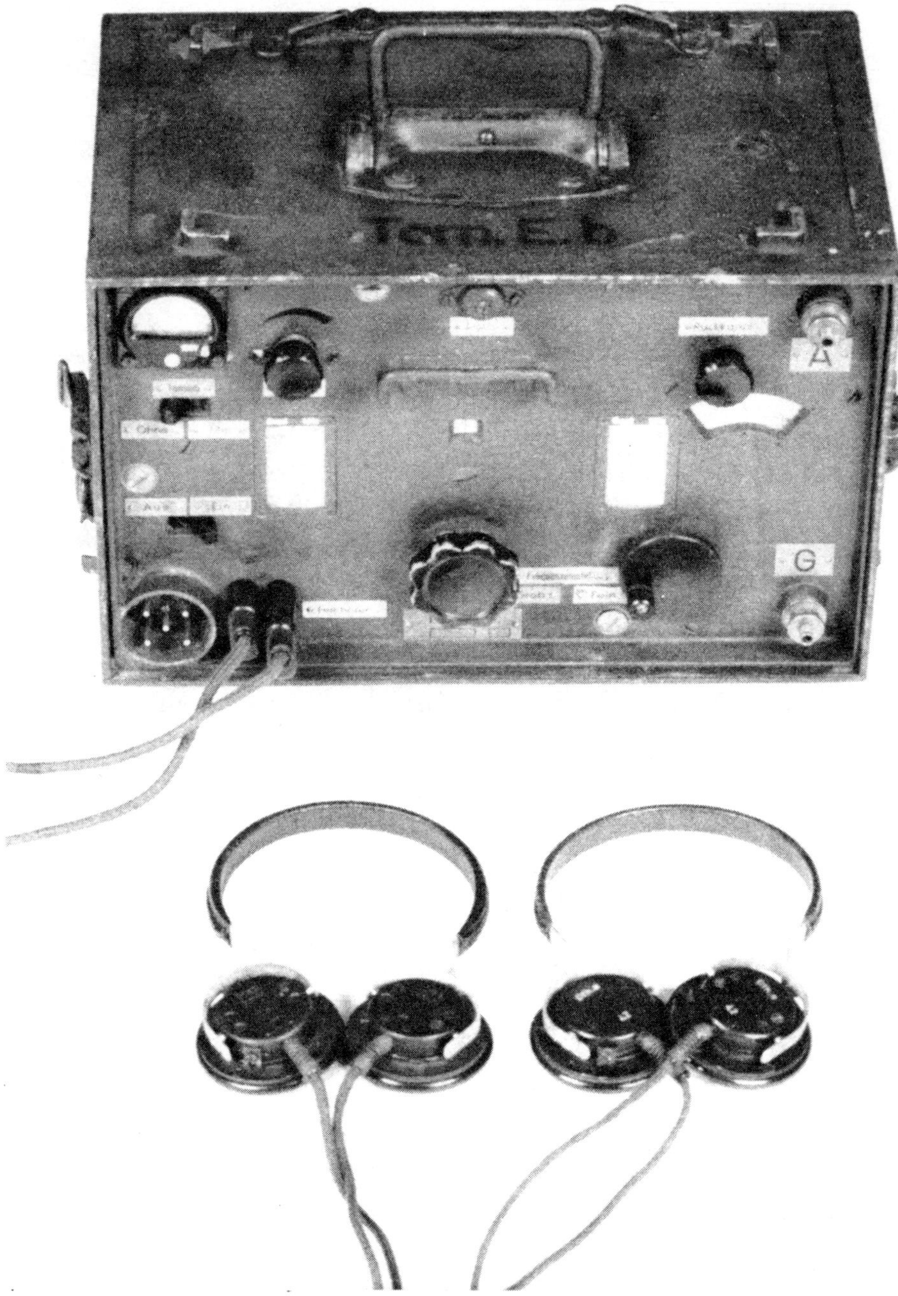

Figure 12. Torn E.B., with headsets installed.

and tone telegraphy; for cw, set this control so that the detector just begins to oscillate, which is usually indicated by a rushing noise in the headset.

NOTE: On some models of the Torn E. b. the **Ruckkoppl** (regeneration control) has a numbered dial to indicate the amount of coupling being used.

(7) To increase selectivity on c-w signals, place the **Tonsieb** (tone-filter) switch on Mit (in). Place switch on **Ohne** (out) for voice or tone reception.

(8) Slowly turn **Frequenzeinstellung Fein** (tuning control) in search of station to be worked, or set **Frequenzeinstellung Fein** on a previous frequency.

(9) Decrease the **Lautst** (volume Control) to desired level.

(10) Upon completion of reception turn the **Aus-Ein** (off-on) switch to **Aus** (off).

Figure 13. The left side of the set, close up view. From top to bottom; voltmeter, lautst (volume control), tonsieb (tone filter switch), tuning dial calibration table and aus-ein off-on switch.

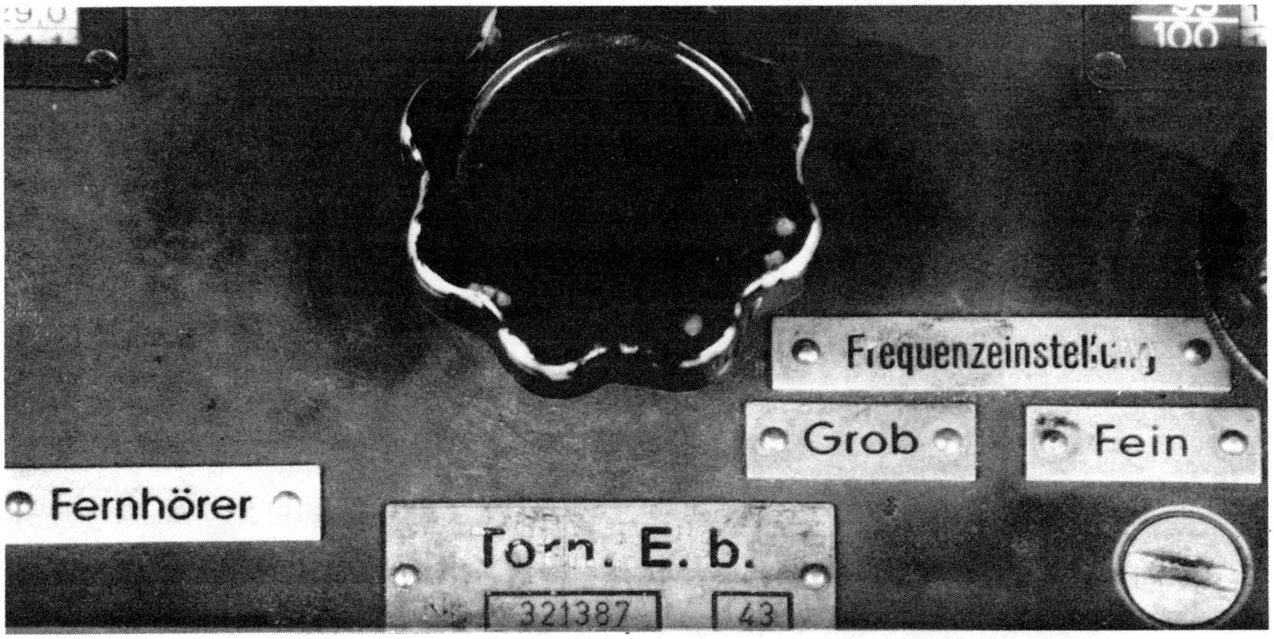

Figure 14. The center lower section of the apparatus case. Band switch, frequenzeinstallung (tuning control) and data plate.

6. FREQUENCY SELECTION:

a. **General.** A scale marked 0 to 100 degrees is fixed to the main tuning capacitor which is operated by the **Frequenzeinstellung Fein** (tuning control) knob. A large disk is attached to a coil drum which is operated by the **Frequenzeinstellung Grob** (band switch) knob, and the frequency corresponding to the capacitor dial scale reading is marked on the drum. Two windows in the panel show the tuning dial calibration tables for each band as the drum is rotated. A third window indicates the number of **kHz** (kc) per division on the tuning dial.

b. **Approximate Interpolation.** It may sometimes be necessary to calculate the frequency of a point between two listed calibrated frequencies. To aid in calculating the proper dial setting or measured frequency corresponding to these intermediate points, the following method should be used. The interpolation described below is the more accurate, but the over-all **kHz** (kilocycles) per **Grad** (dial division) as seen in the band graduation indicator may be used for quick calculations. For example: On band 1, one **Grad** (dial division) is equal to **0.8 kHz** (0.8 kc). In order to calculate the frequency of an incoming signal at a dial setting of 37, proceed as follows:

Dial divisions	Kilocycles	
37	x	unknown frequency
-35	119.4	known frequency
2		

2 (dial divisions) @ 0.8 kHz (kc) = 1.6 kc
Frequency = 119.4 + 1.6 or 121 kc

To calculate the dial setting for 121 kc proceed as follows:

	Dial divisions	Kilocycles
unknown dial setting	x	121.0
known dial setting	35	-119.4
	———	———
		1.6

1.6 kc / 0.8 kHz = dial divisions
Dial setting = 35 + 2, or 37, the dial setting for 121 kc.

To determine the frequency of the dial setting 37 on band 1 more accurately, proceed as follows:

Known dial setting 37. Unknown frequency X.

	Dial divisions	Kilocycles
	40	124.0
	35	-119.4
	———	———
Difference	5	4.6

4.6 / 5 = .92 kc per division
37 - 35 = 2 dial divisions
.92 x 2 = 1.84 kc
Frequency = 1.84 kc + 119.4 or 121.24 kc

The difference between the two calibration chart frequency values is 4.6 kc and corresponds to 5 dial divisions. The frequency in kHz (kc) per dial division is then found by dividing the 4.6 kc by 5 dial divisions, and is 0.92. The difference between the observed dial setting and the calibration chart value dial setting for the lesser frequency is 2 dial divisions. The frequency which this difference represents is then the product of the difference and the kilocycles per division: 0.92 X 2, or 1.84 kc. This is then added to the lesser frequency: 119.4 + 1.84, giving 121.24 kc, the frequency corresponding to the 37 dial setting. Conversely, if it is desired to adjust the receiver to a given frequency between the two calibration chart values, it will again be necessary to make an interpolation to determine the exact dial setting to be used. For example, it is desired that the receiver to adjusted to 121.24 kilocycles. The necessary calculations follow:

Unknown dial setting X. Known frequency 121.24 kc

	Dial divisions	Kilocycles
	40	124.0
	35	-119.4
	———	———
Difference	5	4.6

Difference between known frequency and lesser calibrated frequency = 1.84 kc
5 / 4.6 = 1.09 dial divisions per kc
1.09 x 1.84 = 2 dial divisions
Dial setting = 35 + 2 or 37

The difference between the two calibration chart frequency values is 4.6 kc and corresponds to 5 dial divisions. The number of dial divisions per kilocycles is then found by dividing the 5 divisions by the 4.6 kc. The difference between the desired frequency and the lesser calibration chart frequency is 1.84 kc. The number of dial divisions that this frequency represents is then the product of the difference and the number of dial divisions per kilocycles: 1.09 x 1.84, or 2 dial divisions. This is then added to the lesser calibration chart dial divisions: 35 + 2, giving 37 divisions, the dial setting required for a frequency of 121.24 kc.

7. MAINTENANCE:

a. **Torn. E. b.** Detailed maintenance instructions on this set are not included. The following simple operating precautions should be observed when the set fails to operate:

(1) Check the condition of the batteries with the voltmeter on the panel. Insets which have the voltmeter omitted use an external voltmeter, or if no voltmeter is available substitute known good batteries for those in the receiver. Voltage must lie between the following figures.

Filament voltage .... 1.8 to 2.2 volts
Plate voltage .... 80 to 100 volts

(2) Check all cords, plugs, and connections. Much of the trouble encountered is caused by defective cording and connections.

Figure 15. The Torn. E.b., removed from the case, top-rear view.

Figure 16 a. The Torn E.b., with cover on.

(3) Replace the tubes in order starting with the a-f amplifier and work to the r-f amplifier. To change tubes it is necessary to remove the set from the apparatus case. To remove the set from the apparatus case, loosen the three retaining screws which are on the front panel of the set (figure 10). Figure 15 shows the set removed from the case, and the positions of all the tubes.

b. **Umformersatz EWc.** The servicing of this unit is limited to the replacement of vibrators, fuses, ballast tubes, plugs, and repair of cords and connections.

(1) If the vibrator is working and the set is inoperative, the fuse in the filament circuit may be open, or the ballast tube may be defective. Replace the fuse first, and if the trouble is not fixed, replace the ballast tube.

(2) If set is inoperative and the vibrator is not making a buzzing noise, the vibrator may be defective. Try placing the spare vibrator in the active socket.

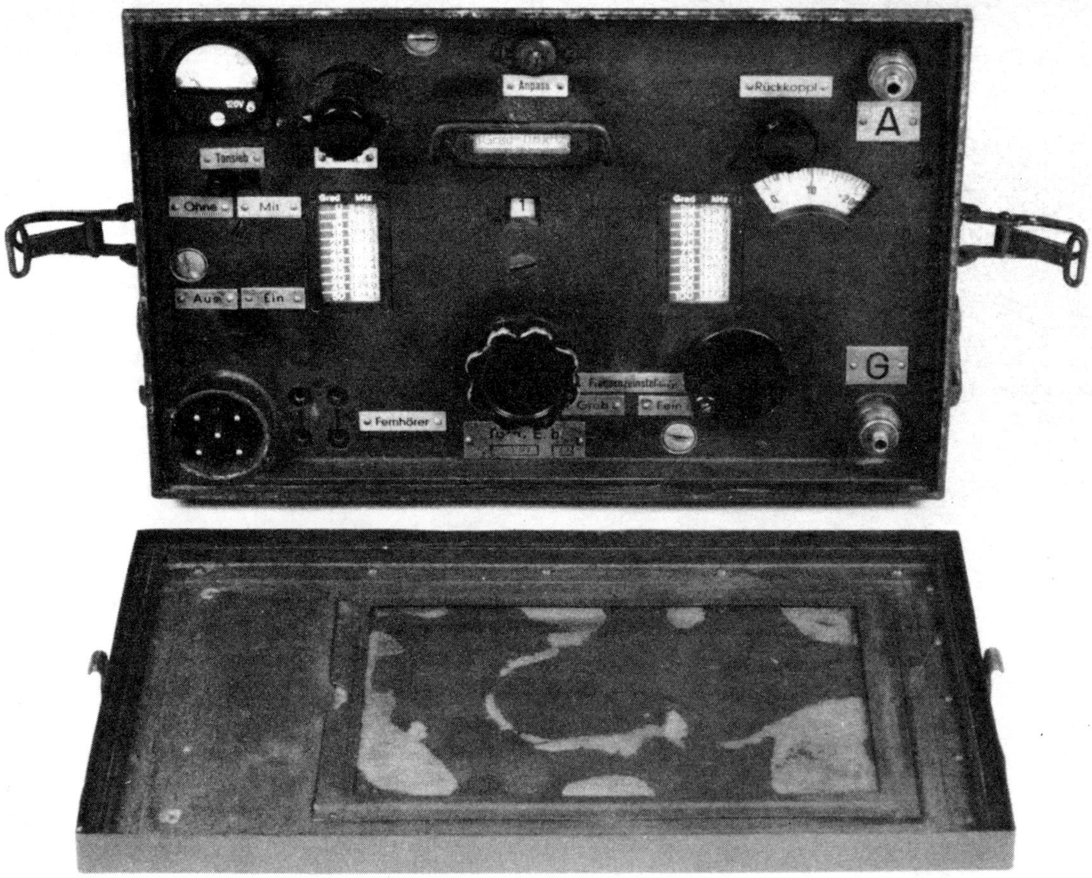

Figure 16 b. The set with the cover removed. The internal frame is for holding instructions and diagrams.

IX

The Kw. E.a - Receiver

The Kw.E.a - Receiver

1. **DESCRIPTION:** The German radio receiver Kurzwellen Empfänger a, abbreviated Kw.E.a is an 11 tube superheterodyne receiver. It is generally used in fixed ground installations, but may be installed in a vehicle. The set is well designed and many of the mechanical parts are cast from a lightweight alloy. Voice, tone, and c-w signals may be received. The sensitivity and selectivity of the set are excellent. The power supply may be either a vibrator unit, an a-c operated rectifier unit, or batteries. The set is housed in a rainproof and dust proof transport case which has a carrying handle on each end. The front cover is held in place by four clasp fasteners. The data plate for the set is shown in figure 1.

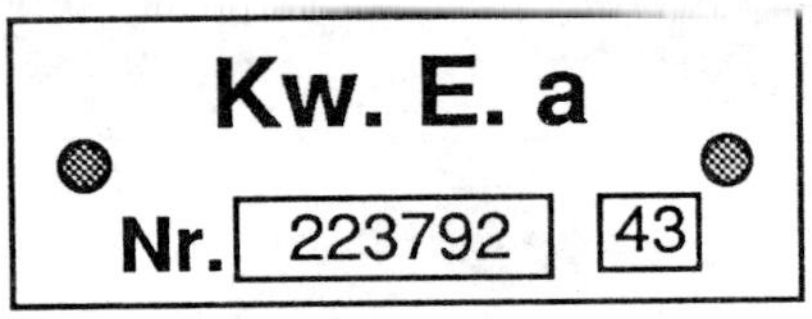

*Figure 1. Nameplate of German radio receiver **Kw.E.a.***

2. **PERFORMANCE DATA:** The performance data and general characteristics of the set are listed below.

PERFORMANCE DATA

Frequency range:	980 to 10,200 kc in 5 overlapping bands
Band I	980 to 1,610 kc
Band II	1,560 to 2,550 kc
Band III	2,470 to 4,060 kc
Band IV	3,940 to 6,395 kc
Band V	6,205 to 10,200 kc
Types of signals which may be received	voice, tone, and c-w
Type of receiver	superheterodyne
Number of tubes	11
2	r-f amplifiers type RV2P800
1	mixertype RV2P800
1	h-f oscillator RV2P800
3	i-f amplifiers RV2P800
1	detector RV2P800
1	a-f amplifier RV2P800
1	avc amplifier RV2P800
1	calibration and beat-frequency oscillator RV2P800
Intermediate frequency	250 kc
H-f oscillator frequency	250 kc above signal frequency
Selectivity	variable
Dial graduations:	
Bands I through IV	10 kc per division
Band V	20 kc per division
Method of calibration	crystal-controlled oscillator
Power requirements:	
Filament	2 v, 2.2 amp, approx
Plate	90 v, 25 ma, approx

Power supply:

Filament	2-volt storage battery, 3-volt dry battery, or rectifier unit
Plate	90-volt battery, vibrator, or rectifier unit

Figure 2. The Kw.E.a. front view with cover removed.

3. **CONTROLS:** A list of the more important controls and their functions follows:

a. The power switch marked **Aus Ein** (off on) turns the filament and plate voltage off or on.

b. The panel meter marked **V** (volts) indicates the filament voltage when the power switch is turned on, and indicates the plate voltage and current when the push buttons on the meter selector panel are pressed.

c. The push button marked **Anoden-Spannung** (plate voltage), when pressed, connects the meter so that it indicates plate voltage. The numbered push buttons marked **Röhren-Strom** (tube current), when pressed, connect the meter so that it indicates the plate current of the correspondingly numbered tube.

d. The switch marked **Selbstreglung ohne mit** (automatic column control out in) switches the automatic-volume-control circuit in or out.

e. The switch marked **Band-breite** (selectivity) is an eight-position selectivity control.

f. The beat-frequency oscillator (BFO) switch marked **Tn Tg** (voice cw) turns on the crystal-controlled BFO in the **Tg** (cw) position.

g. The control marked **Lautst** (volume) is the volume control.

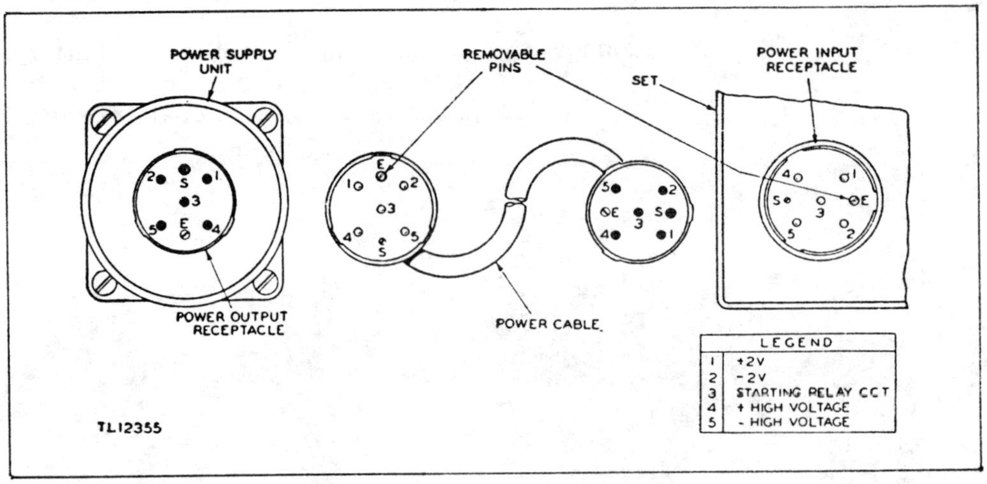

Figure 3. Cording diagram for the Kw.E.a.

h. The control marked **Frequenzeinstellung Grob** (band switch) rotates the turret-mounted r-f coils to change bands.

i. The **Frequenzeinstellung Fein** (fine tuning) control operates the ganged tuning capacitors.

j. The **Stöhöhe** (noise level) control is an r-f gain control.

k. The screwdriver control marked **Anpass.** (antenna matching) is used to match the antenna to the input circuits of the set.

l. The **Vorkreise** (preselector) switch connects an additional tuned circuit in the antenna stage when in position **2**.

m. The **Ankopplung** (antenna coupling) control is a variable capacitor used to obtain the proper coupling between the antenna and the input circuit of the set.

4. POWER SUPPLY INSTALLATION:

a. General.

(1) The power supply usually used with the set in ground installations is an a-c operated rectifier unit. In vehicular installations a type **EWd** vibrator unit which operates from a 12-volt vehicular storage battery is used. For emergency use, a 2-volt storage battery and a 90-volt dry battery can be used.

(2) A cording diagram of the set is shown in figure 3. Make sure that the removable pins are in the **E** positions indicated in this figure, and that the holes marked **S** are open in the power output receptacle and in the receiver end of the power cable plug. Pin **3** of the power input receptacle (which is connected to pin **1** when the power switch is in the **Ein** (on) position) supplies energy for a starting relay when one is used in the power unit.

b. **Rectifier Unit.** If the a-c operated rectifier unit is to be used, place it near the set so that the power cable will reach. Check the power source to make sure its voltage and frequency are the same as those required by the rectifier unit.

c. **Vibrator Unit Type EWd.** When the vibrator unit is used in a vehicular installation, place it near the set so that the power cable will reach, and fasten it securely to the vehicle. Connect the input of the vibrator unit to the vehicular storage battery, and make sure that the polarity is correct and that the connections are clean and tight.

d. **Batteries.** Batteries can be used to supply filament and plate power for emergency operation, and the German batteries should be used when they are available. The following tables list American batteries which can be used to power the set.

FILAMENT SUPPLY (A Battery)

Battery Type	Number Used	Connection	Delivered Voltage	Life (hours)
BA-23	4 6	Series-parallel Series-parallel	3 3	4 15
BA-35	4 6	Series-parallel Series-parallel	3 3	5 14
BA-65	2 4	Series Series-parallel	3 3	5 14
BA-15A	8 12	Series-parallel Series-parallel	3 3	5 14

PLATE SUPPLY (B Battery)

Battery Type	Number Used	Connection	Delivered Voltage	Life (hours)
BA-2	8	Series-parallel	90	30
BA-8	4	Series	90	80
BA-26	2	Series	90	100
BA-36	2	Series	90	30

The life of the batteries is based on continuous operation at a drain of approximately 25 milliamperes from the B batteries and 2.2 amperes from the A batteries. Intermittent operation will greatly increase the life of the batteries. Any 2-volt storage battery may be sued for the filament supply. In the absence of a 2-volt storage battery, a 3-volt dry battery can be used for the filament supply if a resistor of approximately 1/2-ohm is connected in series with one side of the battery. The proper connections of the batteries to the power cable can be determined from figure 3.

5. SET INSTALLATION:

a. **Ground.** In a ground installation, the **Kw.E.a** should be set up and connected as shown in figure 4.

(1) Connect the antenna, which can be any type between 15 and 75 feet in length, to the **A** (antenna) binding post.

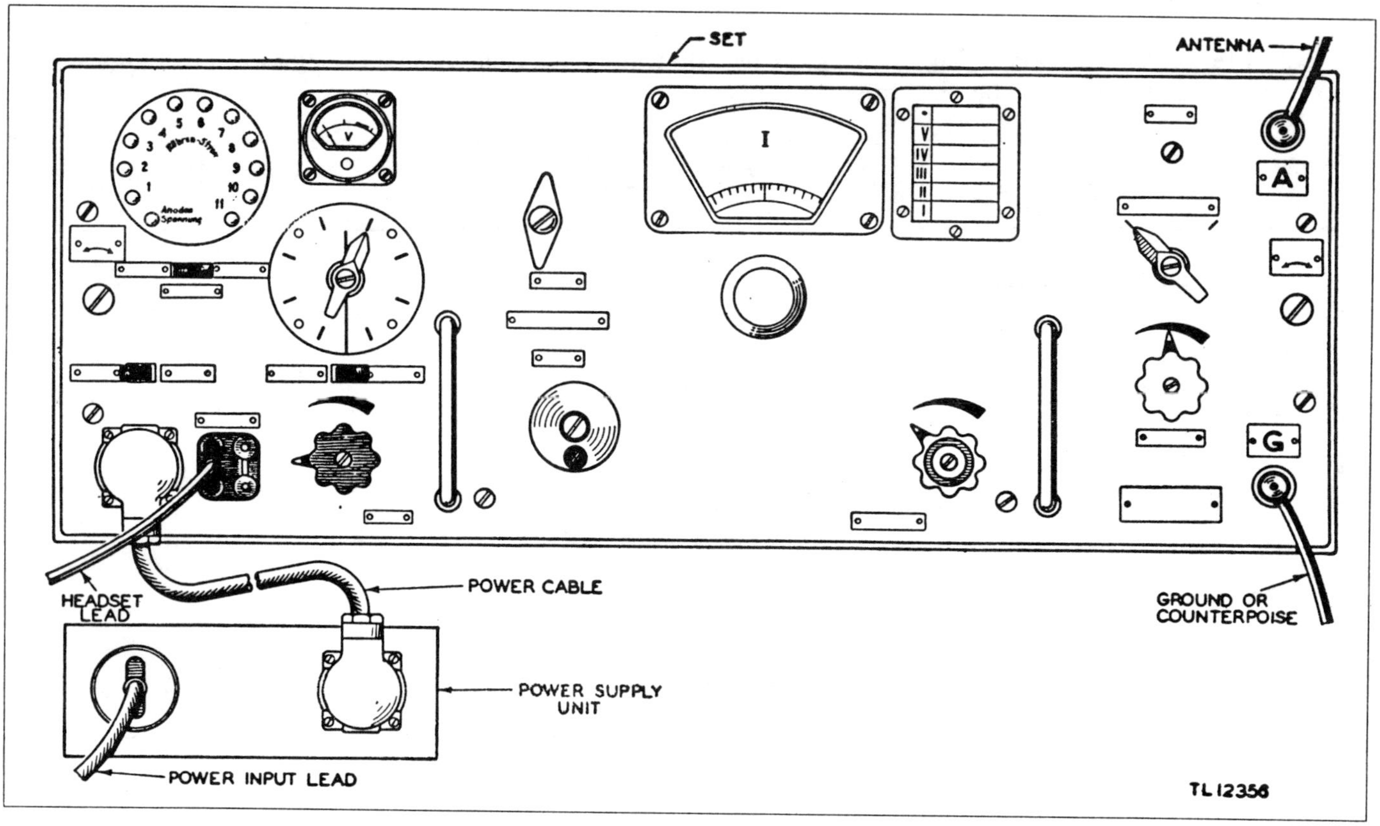

Figure 4. This drawing shows the Kw.E.a. set up for operation.

(2) Connect a ground or counterpoise to the **G** (counterpoise) binding post.

(3) Plug one or two headsets into the jacks marked **Fernhörer** (headset), aligning the plugs with the white marks on the jacks.

(4) Make sure the **Aus Ein** (off on) switch is set to **Aus** (off) and plug the power cable into the power input receptacle on the set and into the power output receptacle on the power unit.

(5) Make a final check of all connections to be sure they are correct before proceeding to operate the set.

b. **Vehicular.** In a vehicular installation, the **KwE.a** should be set up and connected as for a ground installation (figure 4).

(1) Fasten the set firmly to a suitable mounting in the vehicle.

(2) Connect a vehicular antenna to the **A** (antenna) binding post.

(3) Connect the **G** (counterpoise) binding post to the vehicle chassis.

(4) Proceed as directed in subparagraphs a(3), (4), and (5) above.

6. **OPERATION:** After the set has been installed as directed in paragraph **5**, put it in operation as follows:

a. **Antenna Matching.**

(1) Tune in a signal of medium strength on band **V** (subpar. **b** below), and with a coin or screwdriver turn the **Anpass** (matching) control for maximum volume. This matches the antenna to the set for all bands.

(2) If a definite peak is not obtained, but the volume increases up to the right- or left-hand stop, leave the control at the stop concerned. If no change in volume is noticed, leave the control in its center position.

b. Tuning Adjustments for Tone or Voice.

(1) After the antenna has been matched (subpar. **a** above), set the **Frequenzeinstellung Grob** (band switch) control to the desired band, selecting the band from the chart on the front panel (see figure 5). Refer to figure 6 for steps 3-4 and figure 7 for steps 2 and 5-12.

Figure 5. The photograph shows the dial (left) and frequency band chart (right). Each band is color coded.

(2) Set the **Selbstreglung** (avc) switch to the **ohne** (out) position.

(3) Set the **Vorkreise** (preselector) control to position 1.

(4) Turn the **Ankopplung** (antenna coupling) control all the way to the right.

(5) Set the **Band-breite** (selectivity) switch to position **1**, and set the **Tn Tg** (voice cw) switch to the **Tg** (cw) position.

(6) Set the **Aus Ein** (off on) switch to **Ein** (on) and check to see that the panel-meter pointer is within the red portion of the scale.

(7) Press the button marked **Anoden-Spannung** (plate voltage) and check to see that the panel-meter pointer is within the blue portion of the scale.

(8) Turn the **Lautst** (volume) control to the right until background and set noise can be heard in the headset.

(9) Turn the **Frequenzeinstellung Fein** (fine tuning) control until the desired signal is tuned in.

(10) Set the BFO switch to **Tn** (voice), adjust the volume to a comfortable level, and make a final tuning adjustment with the **Frequenzeinstellung Fein** (fine tuning) control. If the received signal is distorted, turn the **Ankopplung** (antenna coupling) control to the left until the distortion is eliminated, and then readjust the volume.

(11) When fading occurs, set the Selbstreglung (avc) switch to the mit (in) position. The noise level can now be adjusted with the Störhöhe (noise level) control. The normal position of this control is at the extreme right, and it should be turned to the left only far enough to reduce any interfering noise to a reasonable level.

(12) When an unwanted signal interferes with the reception of the desired signal, turn the **Ban-breite** (selectivity) switch to position **2, 3,** or **4** until the interfering signal is eliminated. If interference is still present, set the **Vorkreise** (preselector) switch to position **2**, turn the **Ankopplung** (antenna coupling) control slightly to the left, and readjust the volume.

Figure 6. Close-up view of the set from the right side. From top to bottom; antenna matching unit, (anpass) antenna binding post (A) preselect switch, antenna coupling control (ankupplung) ground and data plate.

c. Tuning Adjustments for Cw.

(1) Follow steps (1) through (9) of subparagraph **b** above.

(2) After the signal has been tuned in, set the Band-breite (selectivity) switch to position **5** and tune for the desired pitch of the beat note with the fine-tuning control. Note that as the set is tuned across the transmitted signal, the beat note is much louder on one side of zero beat than

Figure 7. Close-up view of the set from the left side. From top to bottom; meter selector panel meter, avc switch (selbstreglung) selectivity switch (Th-TG), battery plug, headphone jack (fernhorer) and volume control (laust).

on the other side, and in some sets the weaker note cannot be heard at all. Tune to the loudest note in any case.

(3) When fading occurs, set the **Selbstreglung** (avc) switch to the mit (in) position. The noise level can now be adjusted with the **Störhöhe** (noise level) control. The normal position of this control is at the extreme right, and it should be turned to the left only far enough to reduce any interfering noise to a reasonable level.

(4) When an unwanted signal interferes with the reception of the desired signal, turn the **Band-breite** (selectivity) switch to position **6, 7**, or **8** until the interfering signal is eliminated. Note that in position **8** the note is produced on the opposite side of the signal from that produced in position **7**. The BFO frequency, which is crystal controlled at 250 kilocycles in positions **1** through **7**, is changed to 251.8 kilocycles in position **8** to produce the above effect. If interference is still present, set the **Vorkreise** (preselector) switch to position **2**, turn the **Ankopplung** (antenna coupling) control slightly to the left, and readjust the volume. In order to conserve the batteries, set the power switch to **Aus** (off) as soon as reception is complete.

7. SPECIAL CIRCUIT FEATURES: A detailed description of all controls and circuits will not be included in this bulletin, but a few of the controls and their associated circuits peculiar to this set are described in the following paragraphs.

a. The **Band-breite** (selectivity) switch has eight positions; four for use on voice or tone and four for use on cw, although all eight positions can be used on cw. Positions **1** through **4** are used primarily for voice and tone signals, the selectivity increasing in each higher numbered position. The i-f transformers are capacitively coupled, and the selectivity is varied by switching in different values of coupling capacitors. Positions **5** through **8** are used for c-w signals only. A tuned circuit, resonant at approximately 1,000 cycles, is switched into the a-f stage in position **5** and is made progressively sharper in positions **6** and **7**, the i-f circuit remaining as in position **4**. In position **8**, the tuned circuit remains the same as in position **7**, but the crystal-controlled BFO frequency is changed to 251.8 kilocycles to produce a beat note on the opposite side of the received signal. Thus, an interfering signal in position **7** may be eliminated in position **8**. Since the audio circuit is sharply tuned in positions **5** through **8**, these positions should not be used for voice signals.

b. The **Störhöhe** (noise level) control varies the screen-grid voltage of the first and third i-f amplifier tubes, and functions as an r-f gain control to effectively reduce the set and background noise. It operates only when the **Selbstreglung** (avc) switch is in the **mit** (in) position.

c. The **Vorkreise** (preselector) switch, when in position **2**, connects an additional tuned circuit in the antenna stage to provide additional selectivity. This switch should remain in position **1** except when extreme selectivity is required.

d. The **Ankopplung** (antenna coupling) control is a split-stator variable capacitor used to adjust the antenna coupling. It is also used as an attenuation control when very strong signals are being received.

e. The two neon protective lamps are connected in the antenna circuit to protect the coils from any near-by high-power transmitter and from atmospherics.

f. The change-over link should normally connect between terminals 204 and 184. If the set is used for high-speed code reception, this link should be moved to connect between terminals 204 and 183.

8. CALIBRATION: Calibration points are indicated by a red dot on the dial at the following points for each band:

Band I	1,500 kc
Band II	2,500 kc
Band III	4,000 kc
Band IV	6,250 kc
Band V	10,000 kc

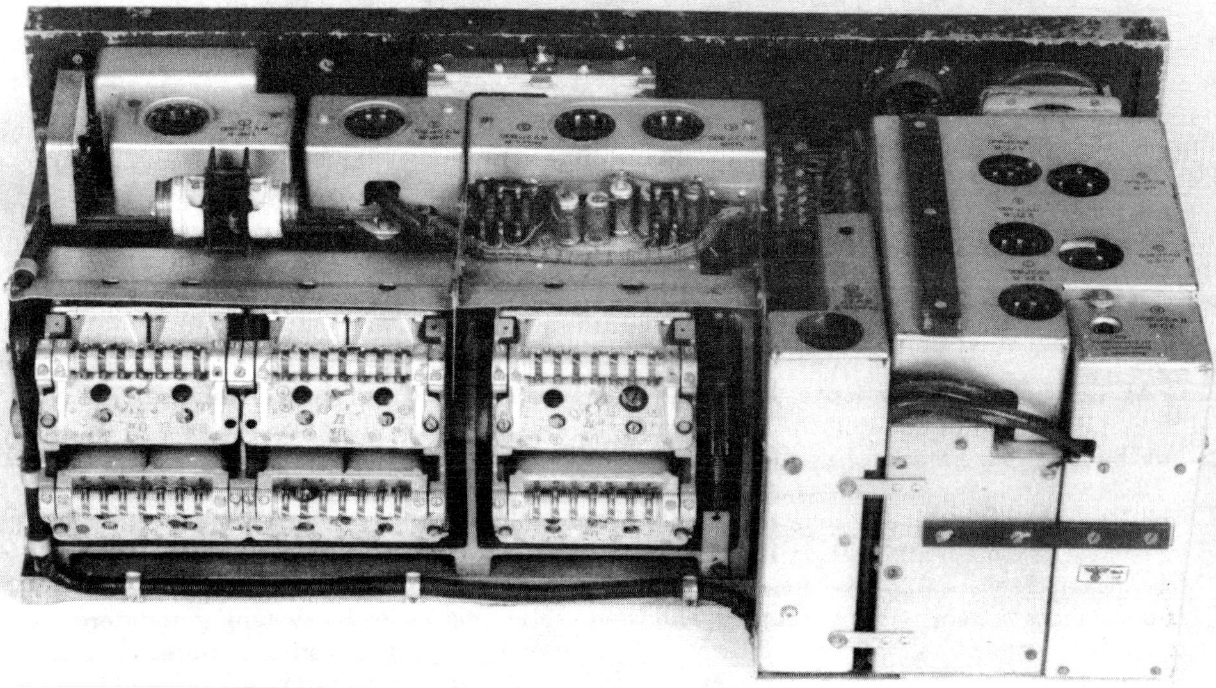

Figure 8. The Kw.E.a. removed from case, top rear view.

To calibrate the set, or to check the calibration for any or all bands, proceed as outlined below.

a. Remove the antenna lead from the set.

b. Use a coin or screwdriver to press and turn the two red-ringed retaining screws one-quarter turn to the left, and slide the set from its case.

c. Set the controls as follows:

(1) **Selbstreglung** (avc) switch to **ohne** (out).

(2) **Band-breite** (selectivity) switch to position **1**.

(3) BFO switch to **Tg** (cw).

(4) **Lautst.** (volume) control all the way to the right.

(5) **Ankopplung** (antenna coupling) control to its central position.

(6) **Frequenzeinstellung Grob** (band switch) to the desired band.

(7) Dial to the red calibration mark.

(8) **Aus Ein** (off on) switch to **Ein** (on).

d. Press the green-ringed calibration button, turn the dial slowly to the right and left of the calibration mark, and listen in the headset for zero beat.

NOTE: When the calibration button is pressed, the output of the 250-kilocycle crystal-controlled calibration oscillator is coupled to the antenna circuit and its harmonics beat with the h-f oscillator to produce the calibration signal.

e. If zero beat does not occur exactly at the calibration mark, set the dial to this point and with a screwdriver carefully adjust the calibration-trimmer capacitor visible through the green-ringed slot (fig. 5) for zero beat.

f. After calibration has been completed, check for other beat notes at 250-kilocycle intervals on each band. If the set is properly aligned, the lowest frequency on each band at which zero beat should occur must be within the following limits:

Band	Frequency	Tolerance
Band I	1,000 kc	+ or - 5 kc
Band II	1,750 kc	+ or - 10 kc
Band III	2,500 kc	+ or - 10 kc
Band IV	4,000 kc	+ or - 15 kc
Band V	6,250 kc	+ or - 30 kc

9. MAINTENANCE:

a. Check all cords, plugs, and connections to make sure they are clean and tight.

b. Set the **Aus Ein** (off on) switch to **Ein** (on) and check the filament and plate supply voltage with the panel meter. (Refer to subpar. **6b** (6) and (7).)

c. Check the plate current of each tube under the following conditions:

(1) Set the **Selbstreglung** (avc) switch to **ohne** (out).

(2) Set the **Band-breite** (selectivity) switch to position **7**.

(3) Set the BFO switch to **Tg** (cw).

(4) Set the **Lautst**. (volume) control all the way to the right.

(5) Remove the antenna lead.

(6) Set the dial to the highest frequency on band **II**.

(7) Press each push button, numbered **1** though **11**, and observe the panel meter indication. The meter pointer should swing above the black mark on the left side of the meter scale. Remove the set from its case (subpar. **8b**) and replace any tube whose indication falls below this mark.

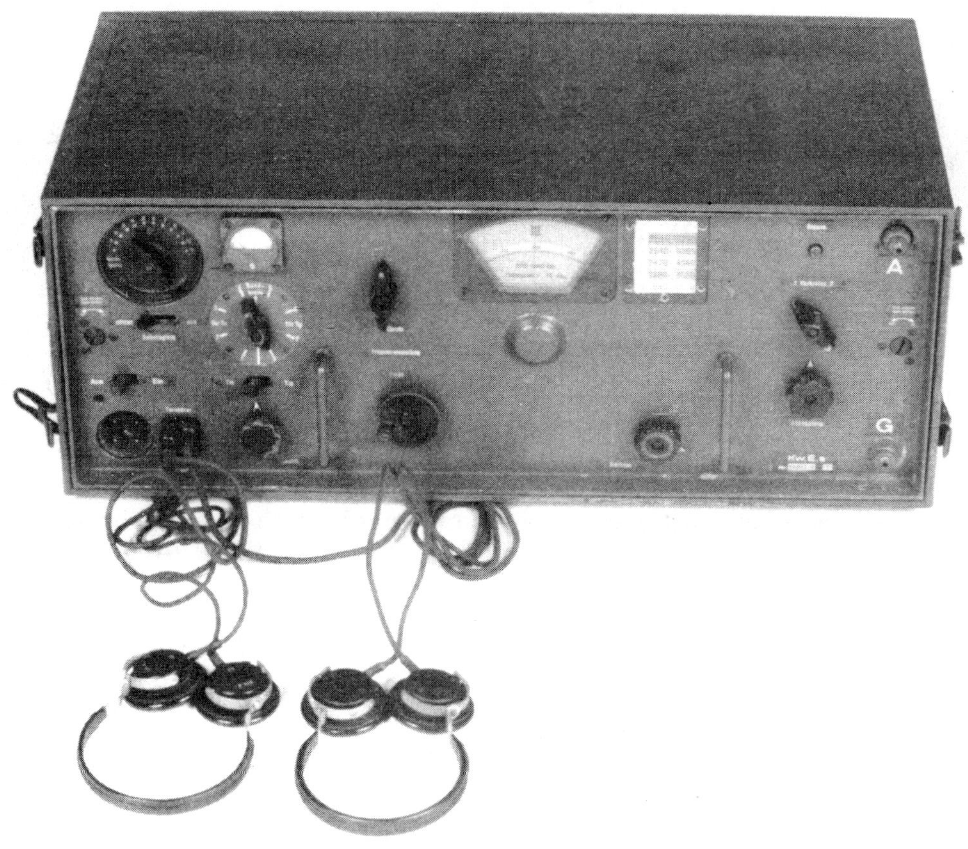

Figure 9. The Kw.E.a. with headphones installed.

X

The Radio Transmitters 10 W.S.c and 10 W.S.h and Radio Receivers Ukw.E.e and Ukw.E.h

The radio operator's position on board the Panzerkampfwagen MkIII. Note the shelf below the 10 watt transmitter for mounting a telegraph key.

The Radio Transmitters
10W.S.c and 10W.S.h
and
The Radio Receivers
Ukw.E.e and Ukw.E.h

1. DESCRIPTION:

a. **General.** Two commonly used German tank sets are the **10W.S.c** (10-watt transmitter, model c)* and the **10W.S.h** (10-watt transmitter, model h). Used in conjuction with the **10W.S.c** is the **Ukw.E.e** (short-wave receiver, model e); used in conjuction with the **10W.S.h** is the **Ukw.E.h** (short-wave receiver, model h). The nameplates are found on each set at the lower center of the instrument panel.

The receivers and transmitters are used in tanks and stationary installations for short range communications with marine stations. The transmitter and receiver chassis are each contained in a steel case, the cover of which clips onto the front panel (figures 1 and 2). There are no lugs or

Figure 1. Nameplate of the 10W.S.c.

projections on this case since provisions for mounting are built in the tanks. On top of each case is a pair of brass strips for grounding purposes. Two cables provide the necessary connections between the transmitter and the receiver. One cable is for sidetone and the other provides for a single antenna to be used for both the receiver and the transmitter. Some models of the set are provided with an interphone system for intercommunication between members of the tank crew. Power is supplied to the receiver and transmitter by the vehicular battery through dynamotors for each unit.

b. **Transmitters.** The radio transmitters **10W.S.c** (10-watt transmitter, model c) and **10W.S.h** (10-watt transmitter, model h) are similar in construction and operation, and differ only in frequency range and dial calibration. The dial of the **10W.S.c** is graduated in kilocycles, while the **10W.S.h** dial is graduated according to wave numbers 50 kc apart, beginning with wave number 241 and progressing to wave number 280. These high-frequency, low-power transmitters are capable of transmitting either tone or voice-modulated signals (a-m). No provisions are made for c-w operation.

Figure 2. Nameplate of the Ukw.E.h.

c. **Receivers.** The radio receivers **Ukw.E.e** (short-wave receiver, model e) and **Ukw.E.h** (short-wave receiver, model h) are seven tube superheterodynes designed for tank use with the transmitters listed above. They are very similar except for frequency range and dial calibration.

d. **Dynamotors.** Two types of dynamotors are used with this equipment: type **U.10al** with the transmitter, and type **E.U.a2** with the receiver. Dynamotors **E.U.a2** and **U.10al** may be used with any radio set with power requirements corresponding to the ratings of these dynamotors. These dynamotors are very similar in construction, and are mounted together with filter chokes and capacitors, in a light-weight cast alloy case. Input voltage is supplied to the dynamotor through a starting relay. This relay is energized when the main switch on the receiver or transmitter with which it is used is turned on. A starting relay is not employed in dynamotor type **E.U.a**, sometimes used in place of **E.U.a2**.

Input voltage is supplied through the power cable when the main switch is turned on. The dynamotors are provided with two connectors: a double-pole input plug, and a five-pole female output receptacle at the opposite end of the case. The double-pole plug is connected through a socket and fuse box to the 12-volt supply. The output receptable connects with the five-pole male plug on one end of the power cable. The three ribs on the output receptacle mesh with the three grooves on the power cable plug. The five contacts of the dynamotor output receptacle are numbered for circuit identification as follows:

1. 12 volts positive
2. 12 volts negative
3. starting relay coil circuit
4. high voltage positive
5. high voltage negative

CAUTION: Use caution in handling this equipment when the transmitter dynamotor is in operation. High voltage (350 v), which is dangerous to human life, is generated by this dynamotor.

Figure 3. U.10a1 dynamotor used with the 10W.S.c and 10W.S.h

Figure 4. E.U.a2 dynamotor used with the Ukw.E.e and Ukw.E.h.

A schematic diagram identifying the various terminals on the dynamotor input and output recaptacles, is located inside the dynamotor cover. The arrow on the dynamotor cover indicates the direction of rotation of the armature.

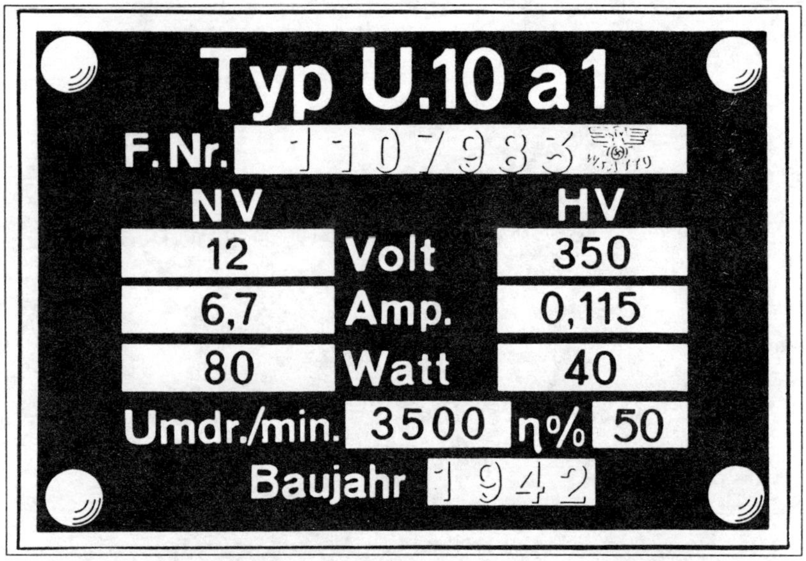

Figure 5. Line drawing showing the dataplate of the U.10al dynamotor.

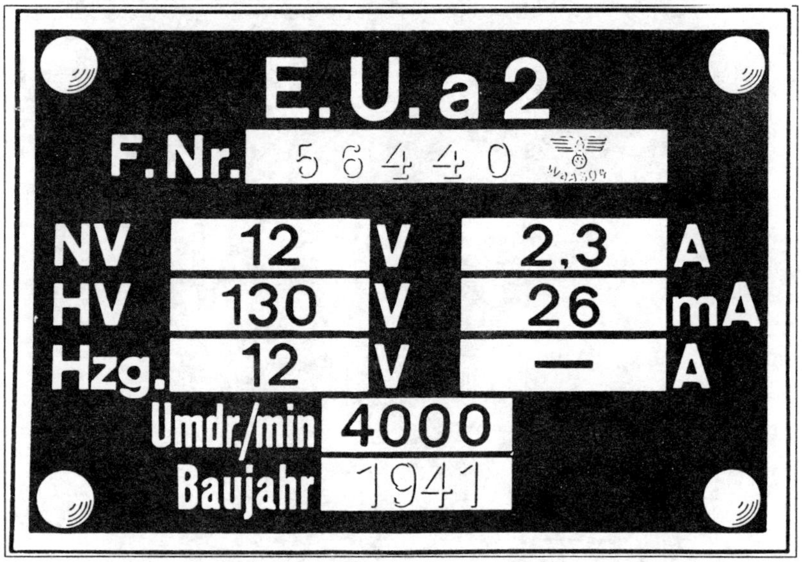

Figure 6. Line drawing showing the dataplate of the E.Ua2 dynamotor.

2. PERFORMANCE DATA: The list below contains the performance data of the sets.

PERFORMANCE DATA

a. Transmitters.

Frequency range:

10W.S.c .... 27.2 to 33.3 mc, in one band

10W.S.h .... 23 to 24.95 mc, in one band

Dial graduation:

10W.S.c .... 50-kc intervals, dial reading x 100=kc

10W.S.h	40 fixed waves numbered from 241 through 280 at 50-kc intervals wave No. 241=23mc wave No. 261=24mc wave No. 280=24.95
Preset frequencies:	2
Transmitter type:	master oscillator-power amplifier (MOPA)
Type of signals emitted:	tone and voice
Distance range:	
Tone:	
Stationary	4 miles
Moving	3 miles
Voice:	
Stationary	3 miles
Moving	1 1/2 miles
Type of modulation:	amplitude
Number of tubes	3
Master oscillator tubetype	RL12P35
Power amplifier tubetype	RL12P35
Modulator and tone	type RV12P4000 oscillator tube
Power output:	6.5 watts (normal) *10 watts (maximum)
Sidetone in the set:	available

*Maximum power is obtained by pressing red push-button switch marked Oberstrich (power increase). This switch is used when aligning antenna, and in emergencies when an increase in power output is desired.

Zero beat receiver with transmitter:	impossible
Method of calibration:	+Use SCR-211-(&)
Antenna:	rod antenna (6 1/2 feet)
Weight:	
Transmitter	19 1/2 lbs
Dynamotor	23 1/2 lbs
Power supply:	12-volt vehicle storage battery through dynamotor German type U.10al
Current drain from 12-volt battery:	
Tone (key down) or voice	7.2 amps (normal) *9.4 amps (maximum)
Tone (key up)	4.5 amps
Stand-by	1.9 amps

Power requirements:
- Filaments and dial lamp .... 12 volts, 2 amps
- Plates .... 350 volts, 100 ma

b. Receivers.

Frequency range:
- Ukw.E.e .... 27.2 to 33.3 mc, in one band
- Ukw.E.h .... 23 to 24.95 mc, in one band

Dial graduation:
- Ukw.E.e .... 50-kc intervals, dial reading x 100=kc
- Ukw.E.h .... 40 fixed waves numbered from 241 through 280 at 50-kc intervals
wave No. 241=23mc
wave No. 261=24mc
wave No. 280=24.95

Preset frequencies: 2

Type of signal which can be received .... tone and voice

Number of tubes: .... 7 (one each as listed)

r-f amplifier	type RV12P4000
mixer	"
h-f oscillator	"
2d i-f amplifier	"
detector	"
a-f amplifier	"

*Maximum power is obtained by pressing red push-button switch marked Oberstrich (power increase). This switch is used when aligning antenna, and in emergencies when an increase in power output is desired.

H-f frequency: .... 3 mc below signal frequency

I-f frequency: .... 3 mc

Power supply: .... 12-volt vehicle storage battery through dynamotor, German type E.U.a or E.U.a.2

Power requirements:
- Plates .... 130 volts, 27 ma
- Filaments .... 12 volts, 1.4 amps
- Dial lamp .... 12 volts, .35 amp
- E.U.a.2 dynamotor relay .... 12 volts, .14 amps

Current drain from 12-volt battery: .... 2.3 amps

Weight:
- Receiver .... 22 lbs
- Dynamotor .... 14 lbs.

3. INSTALLATION:

a. **Location.** Arrange the equipment as shown in figure 5, placing the receiver to the left of the transmitter for ease in making connections. When receiver and transmitter racks are provided in the vehicle, slide the units into the racks and press down the clamps on each side. If dynamotor racks are available, place the units in the racks and tighten the wingnuts on the bolts which fit into the slotted lugs at the output end of the case. If racks or mountings are not available, firm supports for the equipment will have to be constructed.

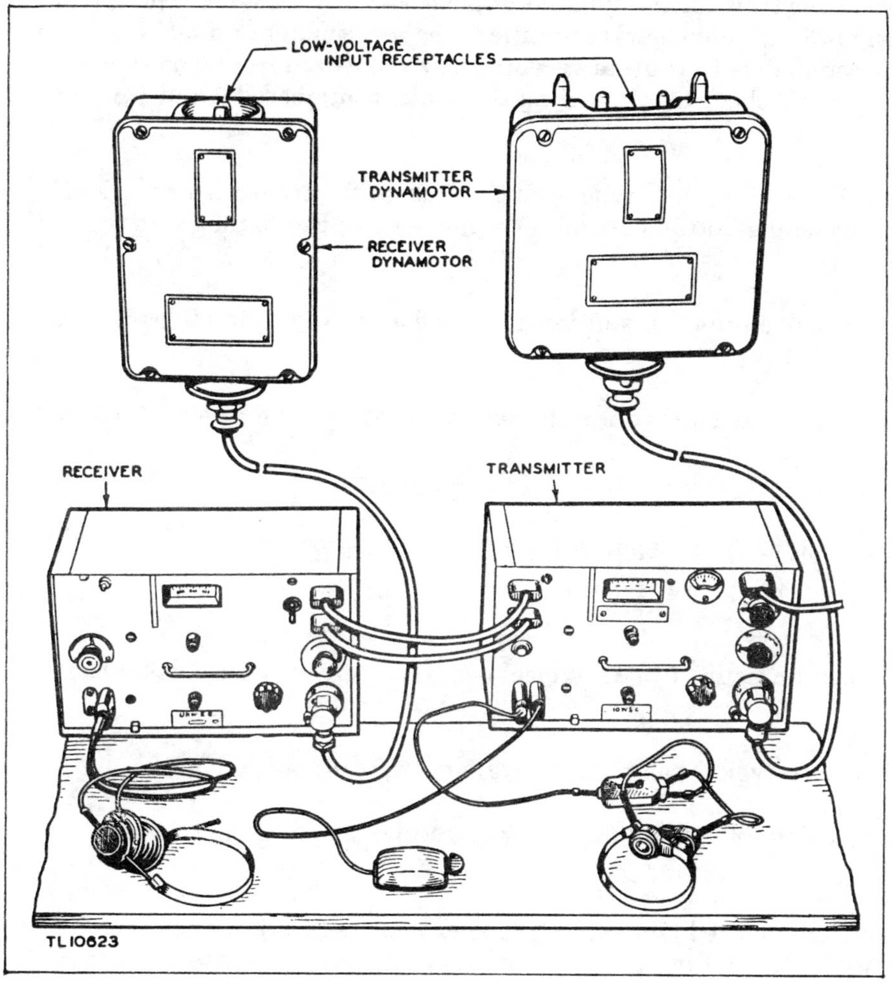

CAUTION: Before making any connections, make sure that the control on the receiver marked Aus Ein Lautstarke (off-on-volume) and the transmitter control marked Aus Tn Empfang Tg Tonend (off-voice-receiver-tone) are in the Aus (off) position. In making the following connections, be certain to observe polarity; that is, in each case the larger terminal of each jack is connected to the larger terminal of the corresponding jack.

Figure 7. Line drawing showing installation of the 10W.S.c and Ukw.E.e with dynamotors and accessories.

b. **Cabling.**

(1) Connect a cable from the jack marked **G.** (counterpoise) **Ant.** (antenna) on the receiver, to the jack marked **Ant.** (antenna) **Empf.** (receiver) **G.** (counterpoise) on the transmitter.
(2) Connect the sidetone cable from the jack marked **z. Sender** (to transmitter) on the receiver to the jack on the transmitter marked **z. Empf.** (to receiver).

(3) Connect the receiver and transmitter to their respective dynamotors, first checking to see that the screws and removable pins are in the correct positions (fig. 6). Observe that the power cables are identical in the construction; the unit with which either is used is determined by the location of the pins and screws in the plugs and receptacles. Figure 6-1 shows the cording for the transmitter. The removable pins in the male receptacles used in the transmitter cording

are placed in the holes marked **S**, and in the female receptacles the screws are likewise placed in the holes marked **S**. In the receiver cording (fig. 6-2) the removable pins in the male receptacles are placed in the holes marked **E**, and the screws in the female receptacles are placed in the holes marked **E**.

(4) When the dynamotors are mounted in their respective racks, connection to the 12-volt vehicular battery is supplied through the terminals on the dynamotor case and junction box socket on the rack. If the mounting racks for the dynamotors are not available, direct connection to the 12-volt vehicular battery can be made at the terminals on the dynamotor case. In this case a 15-ampere fuse will have to be used in the receiver dynamotor positive lead, and a 40-ampere fuse in the transmitter dynamotor positive lead. Observe input polarity. If the dynamotor input terminals are not clearly identified, use a voltmeter and make sure the voltage and polarity on the contacts of the output receptacle are shown in legend on figure 6. The dynamotor can be run for this check by connecting the contact numbered 3 and 1 with a jumper.

(5) Plug a microphone, either the throat or hand type, into the jack on the transmitter marked **Mikrofon** (microphone). Any single-button carbon microphone equipped to plug into the jack will work satisfactorily.

(6) Plug a telegraph key into the jack on the transmitter marked **Taste** (key). Any telegraph key will work.

(7) Plug a headset into either of the jacks marked **fernhorer** (headsets) on the receiver. The set is now ready for operation.

4. **OPERATION:** After the set is installed (par. 3), it is put into operation as follows:

a. **Transmitter Operation.**

(1) Turn the main switch as shown in figure 7 to **Tn** (voice) position. Allow 60 seconds for the tubes to warm up.

NOTE: When the main switch is in the **Tn** (voice) position, the transmitter carrier is on.

(2) Select the desired frequency using the control marked **Frequenz-Einstellung** (tuning control).

(3) Pressing the push-button switch marked **Oberstrich** (power increase), adjust the control marked **Ant. -Abstimmung** (antenna tuning) for maximum reading on the antenna meter.

(4) Press the microphone switch, usually marked **Drucken** (press), and speak into the microphone. Deflections on the antenna meter indicate modulation.

(5) To operate the transmitter on tone, place the main switch to **Tg tonend** (tone) position, and use the telegraph key.

b. **Receiver Operation.** Refer to figure 8.

(1) Turn the main switch on the transmitter to the **Emfang** (receive) position.

(2) If the transmitter with which the set is in communication causes the receiver to block, turn the **Fern-Nah** (far-near) switch on the receiver to the **Nah** (near) position. This reduces the coupling to the receiver antenna input circuits, thus protecting these circuits when operating in a strong r-f field.

(3) Turn the receiver main switch marked **Aus Ein Lautstarke** (off-on-volume) to **Ein** (on). Allow 60 seconds for the tubes to warm up.

(4) The pointer of the control marked **Feineinstellung** (fine tuning) should be at **0**.

(5) Tune to the desired frequency using the control marked **Frequenz-Einstellung** (tuning control).

(6) Control the volume of the signal in the headset with the control marked **Aus Ein Lautstarke** (off-on-volume).

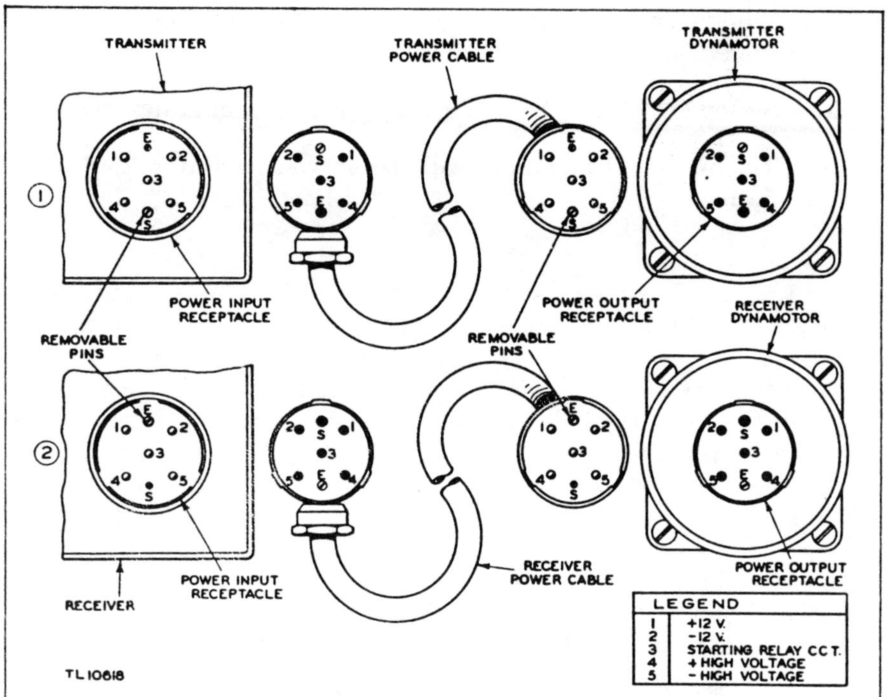

① Transmitter cording diagram.
② Receiver cording diagram.

Figure 8. Cording diagram for transmitters 10W.S.c and 10W.S.h and receivers Ukw.E.e and Ukw.E.h.

(7) The control marked **Feineinstellung** (fine tuning) provides a fine adjustment of the h-f oscillator frequency, which permits following a drifting signal.

(8) The receiver may be operated independently of the transmitter antenna by connecting an antenna to the jack on the receiver marked **Ant.** (antenna) and grounding the jack marked **G.** (counterpoise). The switch marked **Fern** (far) and **Nah** (near) should be in the **Nah** (near) position every time the transmitter carrier is on.

5. PRESET CHANNEL ADJUSTMENTS: To preset frequency channels of either the transmitter or receiver proceed as follows:

a. Be sure that both preset-channel locks (figs. 7 and 8) are in their normal position, turned to the extreme right, and released. In the case of the receiver, the control marked **Feineinstellung** (fine tuning) must be turned so that the pointer is at **0**.

b. Turn the control marked **Frequenz-Einstellung** (tuning control) until the desired channel number, either **I** or **II**, appears in one of the preset-channel windows (figs. 9 and 10). The number **I** appears in the upper window and the number **II** appears in the lower preset-channel window.

c. With a screwdriver or coin, press in the corresponding preset-channel lock, and turn it to the left as far as it will turn.

d. Turn the control marked **Frequenz-Einstellung** (tuning control) until the dial indicates the required frequency.

e. Turn the preset-channel lock to the extreme right and release.

f. Whenever the channel number preset appears in its window, the transmitter or receiver is tuned to the frequency originally chosen.

g. Set up the other preset channel in the same manner.

h. The preset arrangement will also serve as a dial lock, although it is very easy to change frequency if desired.

6. **CALIBRATION:** Check the calibration of the **10W.S.c** and the **Ukw.E.e** at a frequency near 30 megacycles, and the calibration of the **10W.S.h** and the **Ukw. E.h** at a frequency near 24 megacycles, by means of a Frequency Meter Set. Since a strong response normally cannot be obtained from the frequency meter at these frequencies, dial settings are not included in the calibration book used with the instrument. However, by adjusting the meter to 3 mc and using its 10th harmonic in checking the **10W.S.c** and the **Ukw.E.e**, and its 8th harmonic in the checking **10W.S.h** and the **Ukw.E.h**, a signal of sufficient strength to check the receiver or transmitter frequency may be obtained. If zero beat is not obtained at the tested frequency, calibration is necessary. To calibrate the transmitter loosen the screw in the right of the calibration adjustment cover, slide back the cover, and with a screwdriver adjust the calibration screw until zero beat is obtained. In calibrating the receiver, hold the pointer of the **Feineinstellung** (fine tuning) control at **0**, and with screwdriver adjust the screw-ended shaft through the center of the knob, turning the shaft until zero beat is obtained.

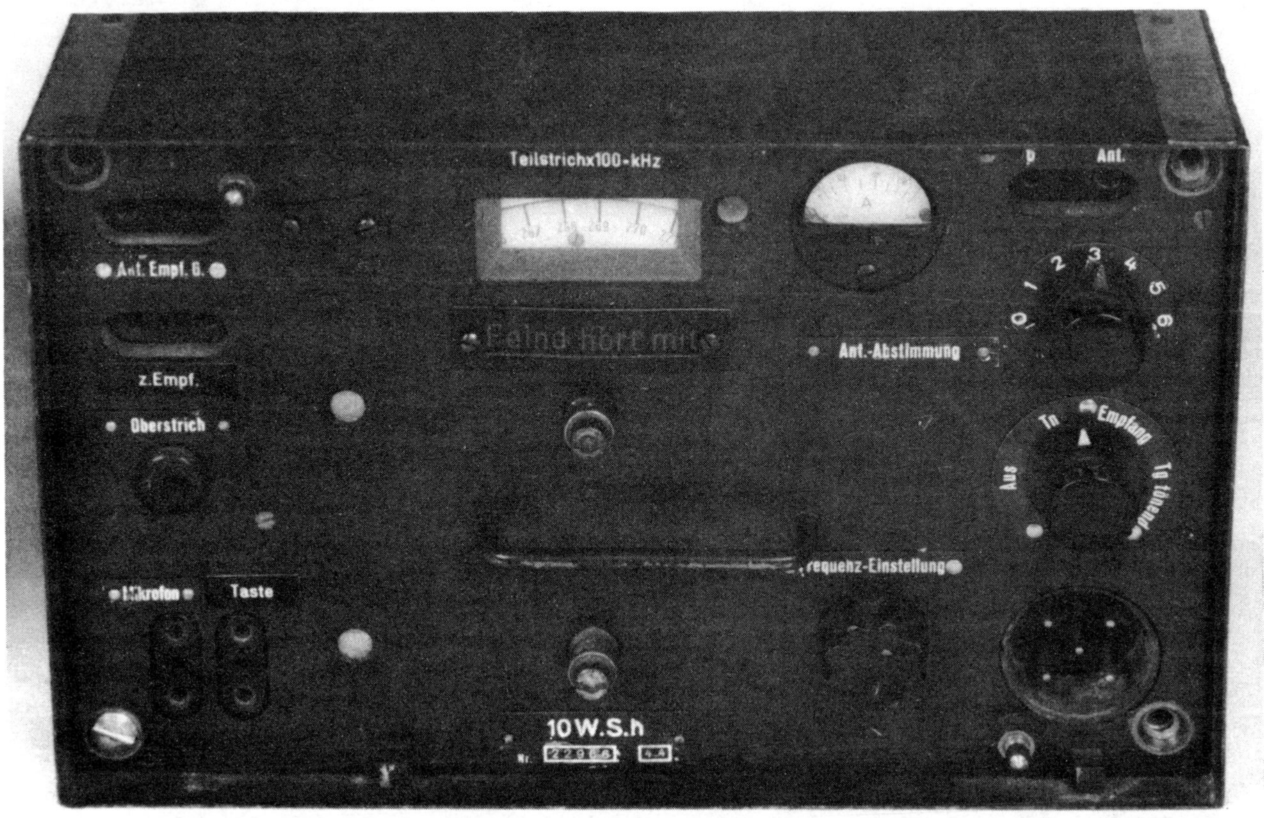

Figure 9. Front view of the 10W.S.c transmitter.

7. **MAINTENANCE:** If the set fails to operate, follow the general instructions given below:

a. Check all cords, plugs, and connections. Much of the trouble encountered will be in cording and connections. If the dial lamp lights, filament voltage is entering the set. If the neon lamp glows, high voltage is present.

b. Replace the tubes in order, starting with the output and progressing toward the master oscillator in the transmitter. In the receiver, start with the a-f amplifier and work to the r-f amplifier. To change the tubes, it is necessarey to remove the set from its case. Do this by removing the four retaining screws on the front panel of the set (figs. 9 and 10). Each retaining screw is surrounded by a red ring. Then remove the set from the case. Figures 11 and 12 give a top rear view of the sets with the case removed, showing the position and function of all the tubes.

NOTE: The same tube in the transmitter serves as modulator and tone-oscillator. Therefore, if tone operations is satisfactory but voice modulation is impossible, look for trouble in the microphone circuit rather than in this tube.

c. To remove either the power amplifier or master oscillator tube from its socket:

(1) Loosen the thumbnuts and remove top-cap leads.

NOTE: On the German tube type RL12P35, only the two top-caps, which have thumbnuts, connect to internal elements of the tube. A third cap is found on some types of this tube, which serves no purpose other than covering the seal-off tip.

(2) Grasp the tube, and twist to left until free, then lift it out of the socket. Observe the relative position of the guide pin on the side of the tube base, since this will simplify replacing the tube.

NOTE: When the chassis is replaced in the case, all tubes should be seated firmly in their sockets. This procedure will reduce the possibility of a tube jarring from its socket, causing set failure as well as difficulty in removing the chassis from the case.

Figure 10. Front view of the Ukw.E.h receiver.

Figure 11. Rear view of the 10W.S.c removed from the case.

Figure 12. Rear view of the Ukw.E.h removed from the case.

Figure 13. The 10W.S.c with front cover in place.

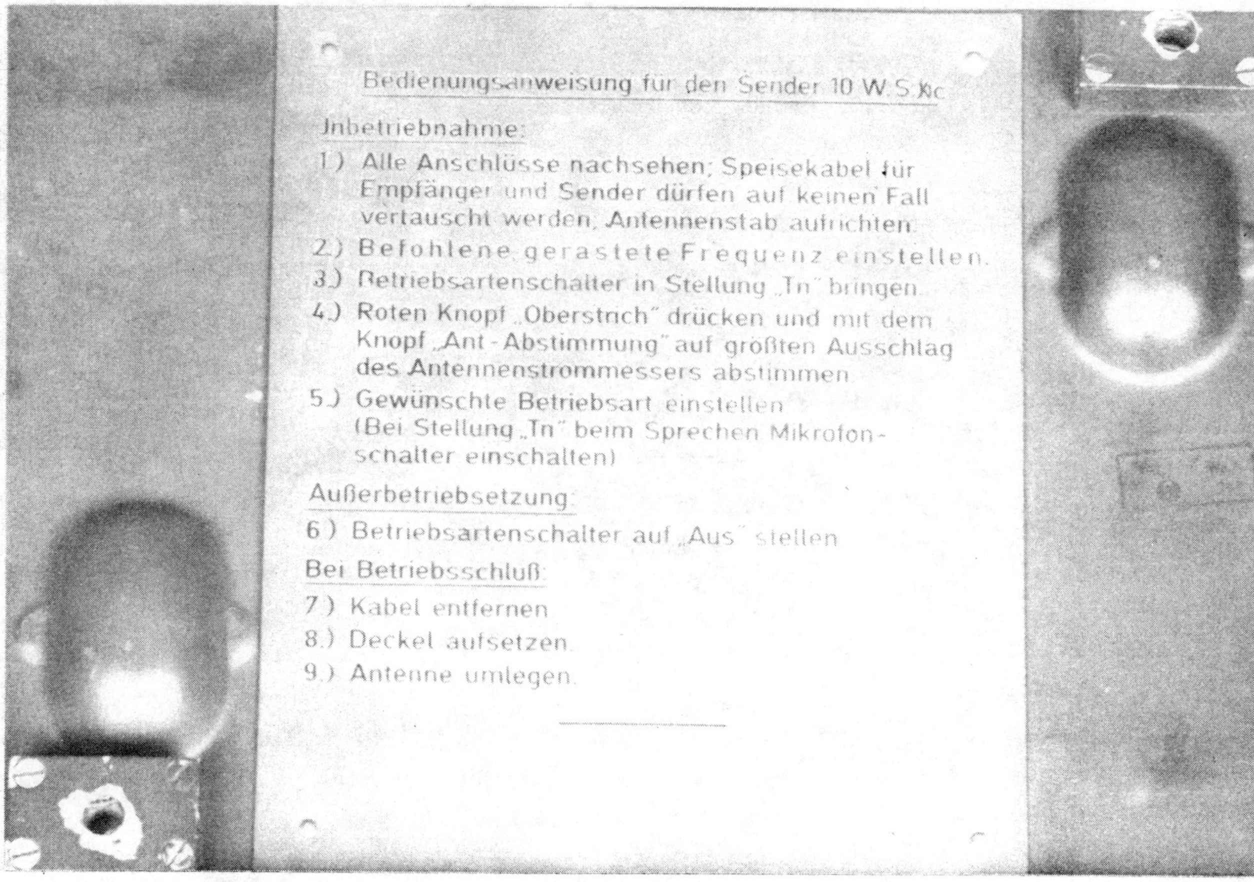

Figure 14. Inside view of the front cover for the 10W.S.c.

Figure 15. This illustration shows the tank intercom box for crew members with the headphones and throat microphone installed.

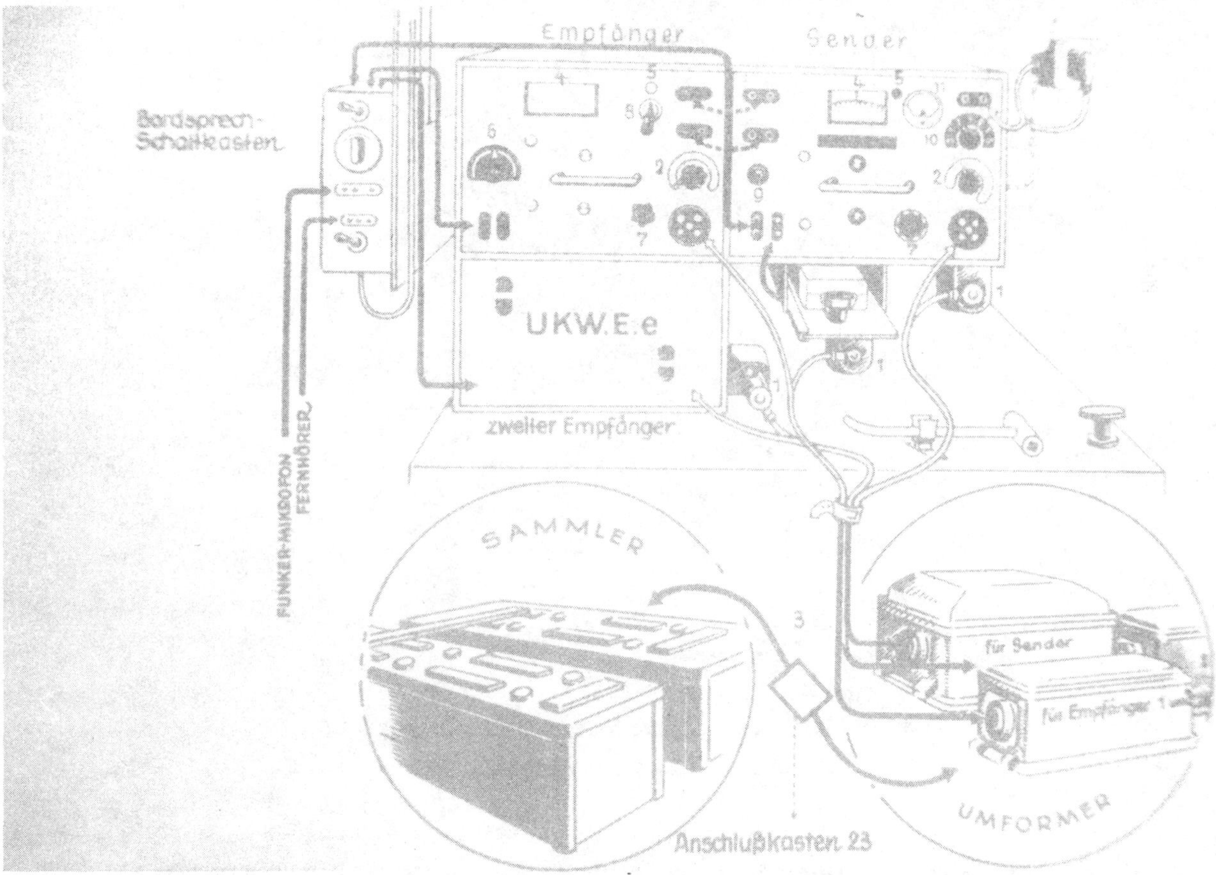

Figure 16. This illustration, taken from page 23 of the Manual D656/27 Tiger Fibel, shows the installation of the radio system on board a Tiger tank.

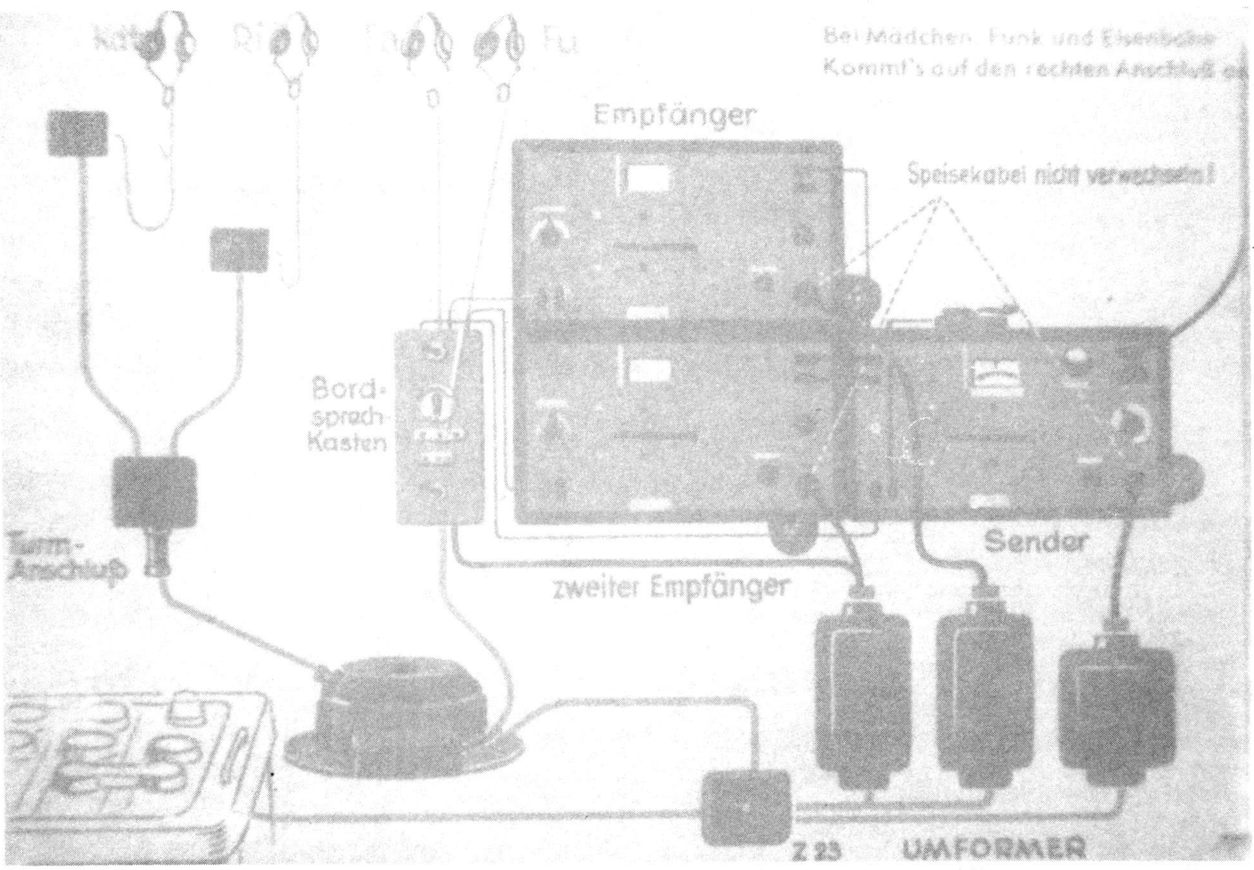

Figure 17. Another schematic from D655/27 Panther Fibel. This illustration further shows the intercom system. The use of a second receiver would only be found on Command and Control vehicles such as the platoon leader and company commander's tank.

Figure 18. The dynamotors E.U.a2 and U.10a1 mounted in their racks on board the Panzerkampfwagen MkIII.

XI

The Accessory Items

Figure 1. This illustration shows two examples of the hand microphone. Note the example at left with the operational security warning label, Feind hort mit! (the enemy is listening).

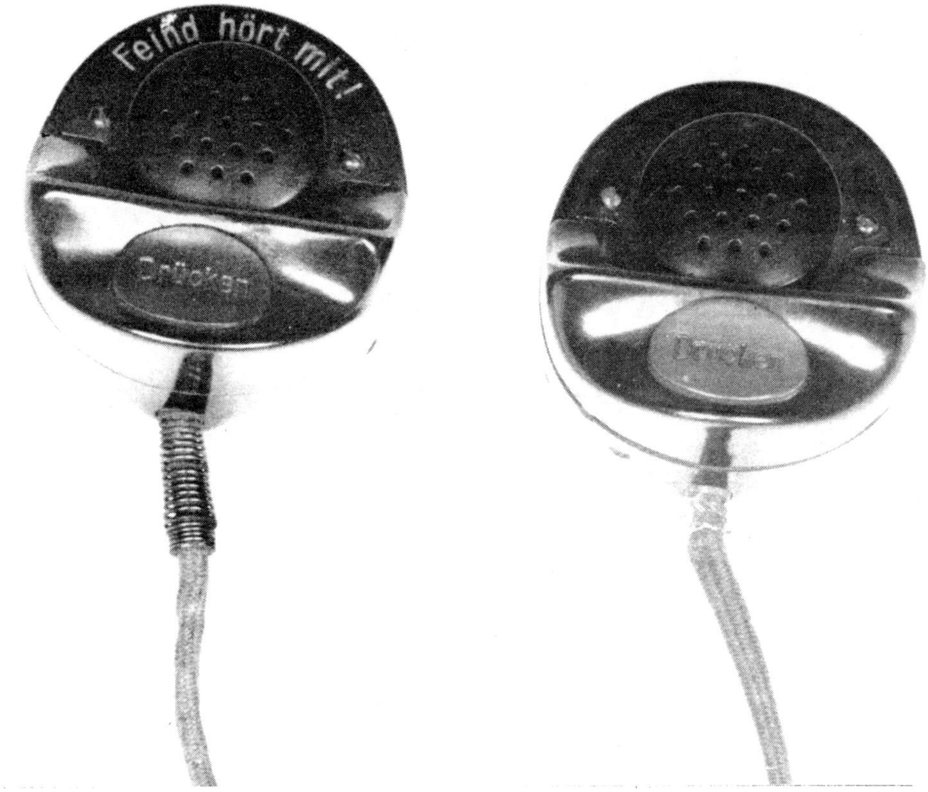

Figure 2. Another view of the hand microphones showing the reverse side at right. To use, press the button labeled Drücken (push) and speak into the microphone.

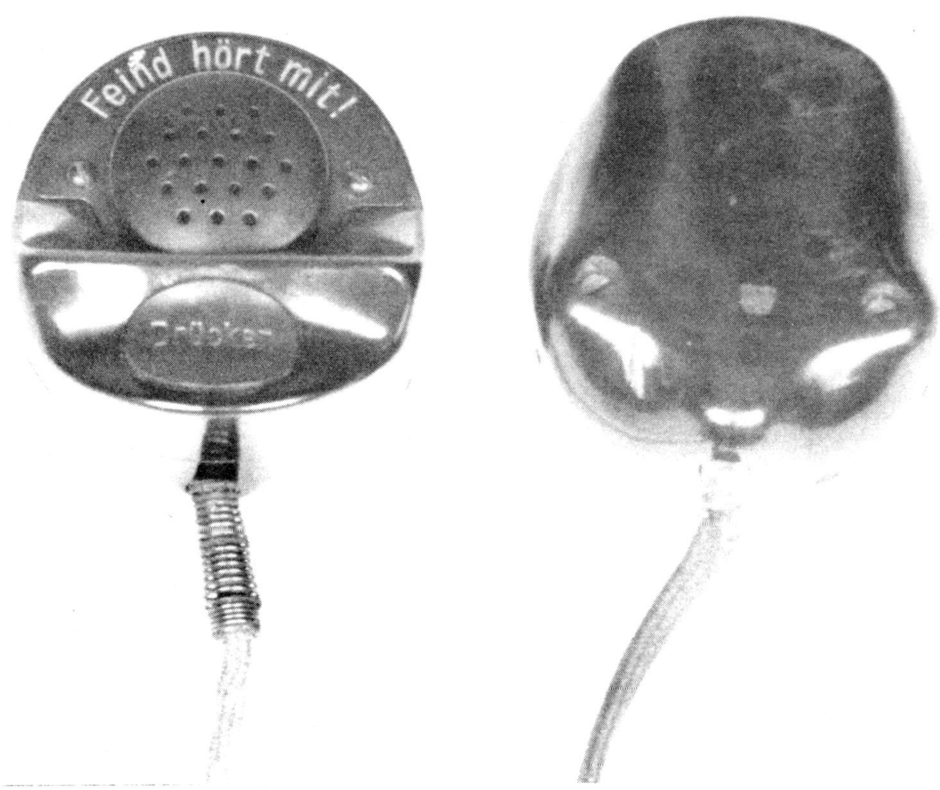

Vorschriftsmäßiges Halten des Handmikrofons.

Figure 3. Extract from an original German training manual showing how to use the hand microphone.

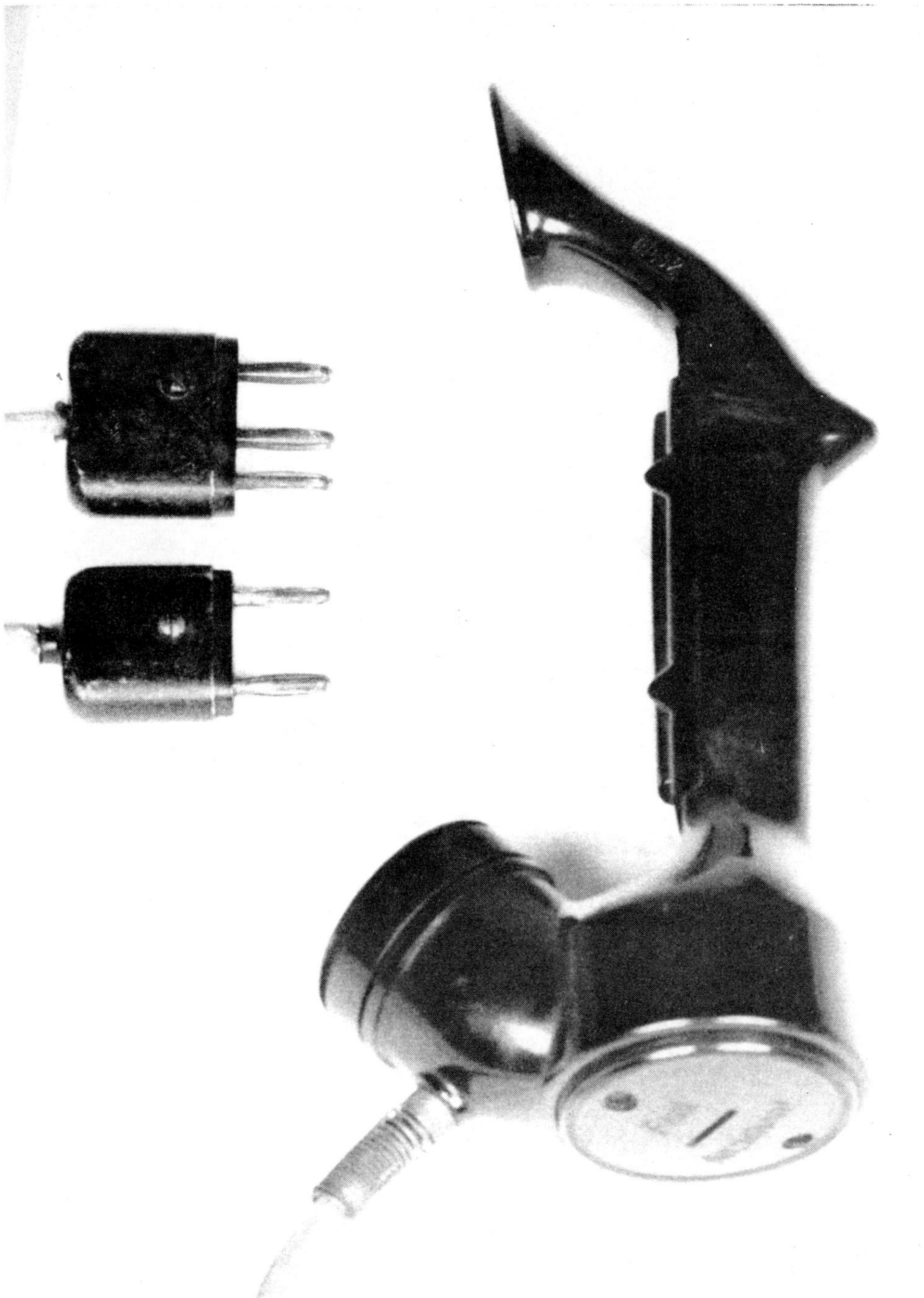

Figure 4. This illustration shows the German handset for radios. It is designed specifically to be used while wearing a steel helmet. The headphone is just large enough to cover the ear. Note the plug ends at the left.

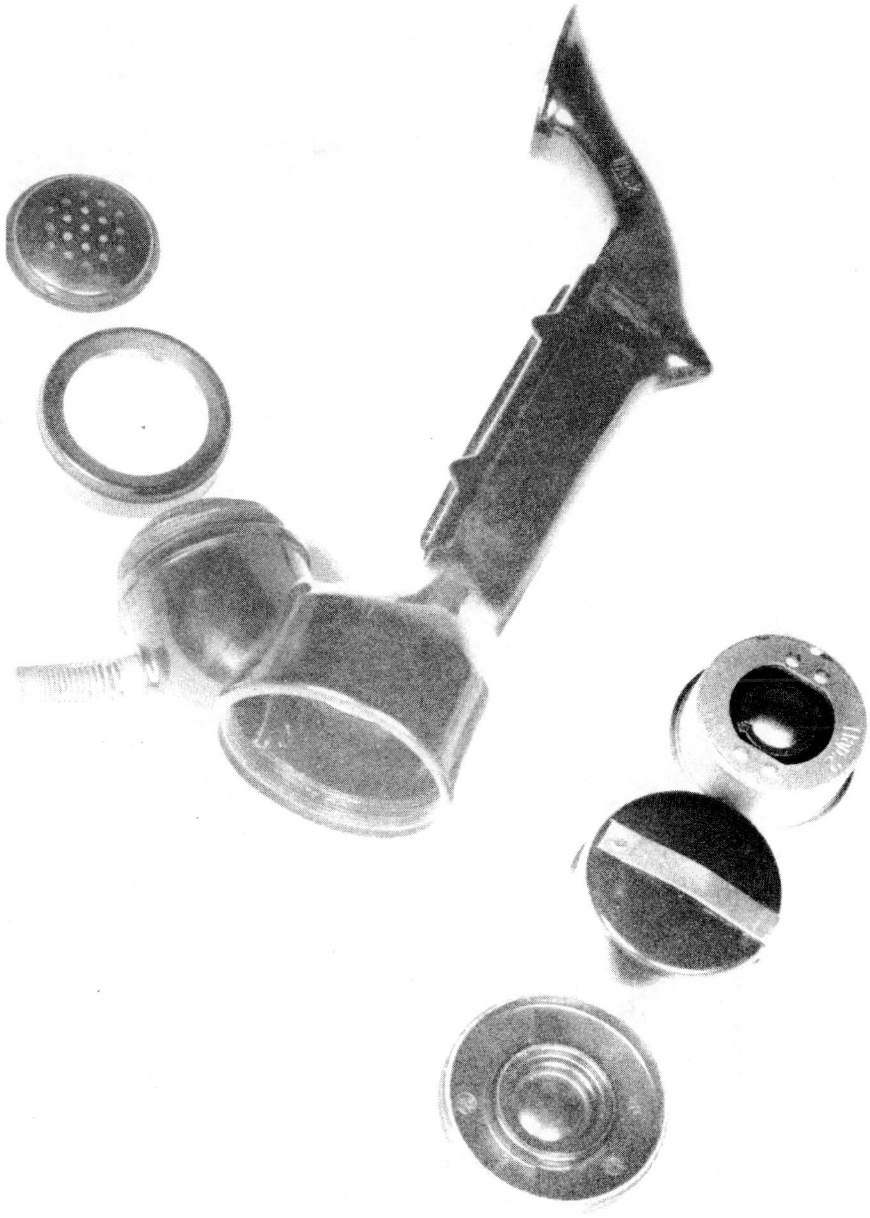

Figure 5. This view shows the handapparate 2, or Hap 2, with internal components removed.

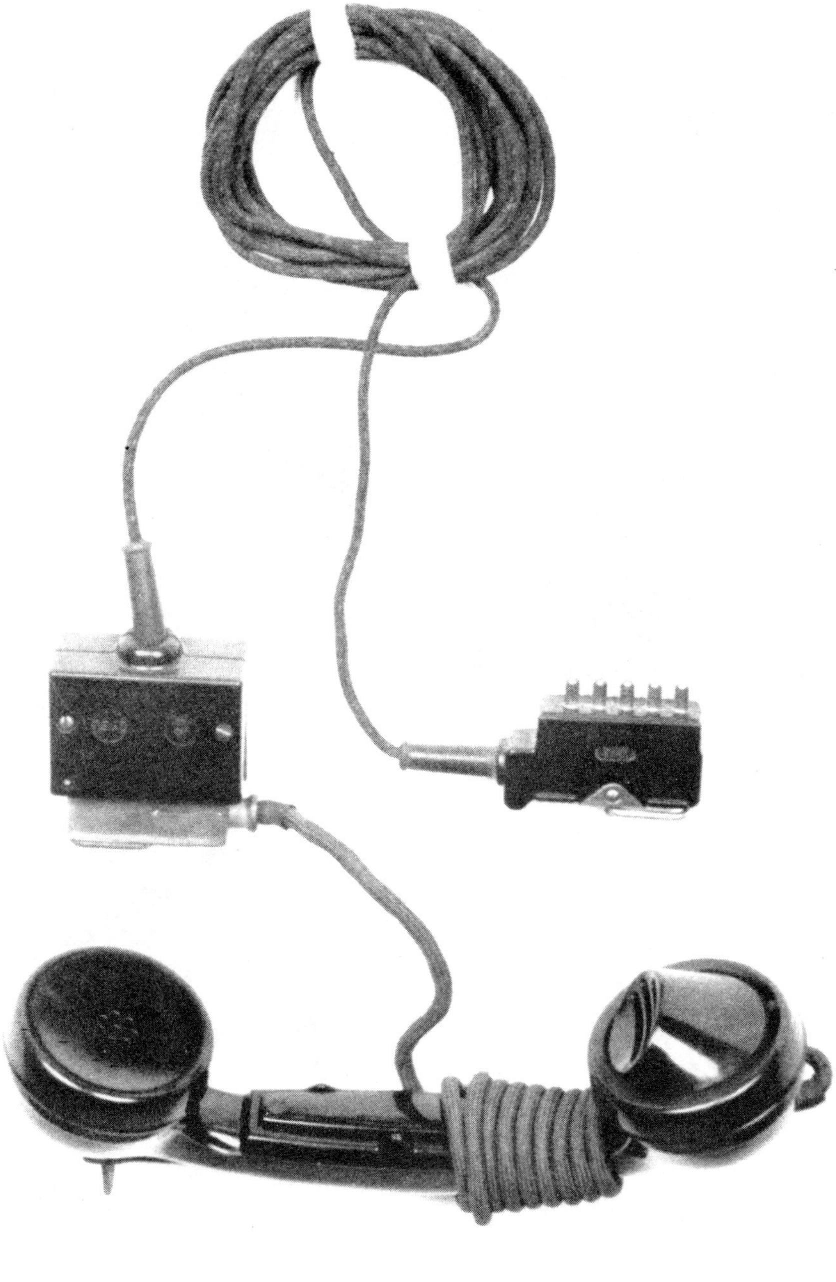

Figure 6. Handset used with the German field telephone. Note the extension cord in the center of the illustration.

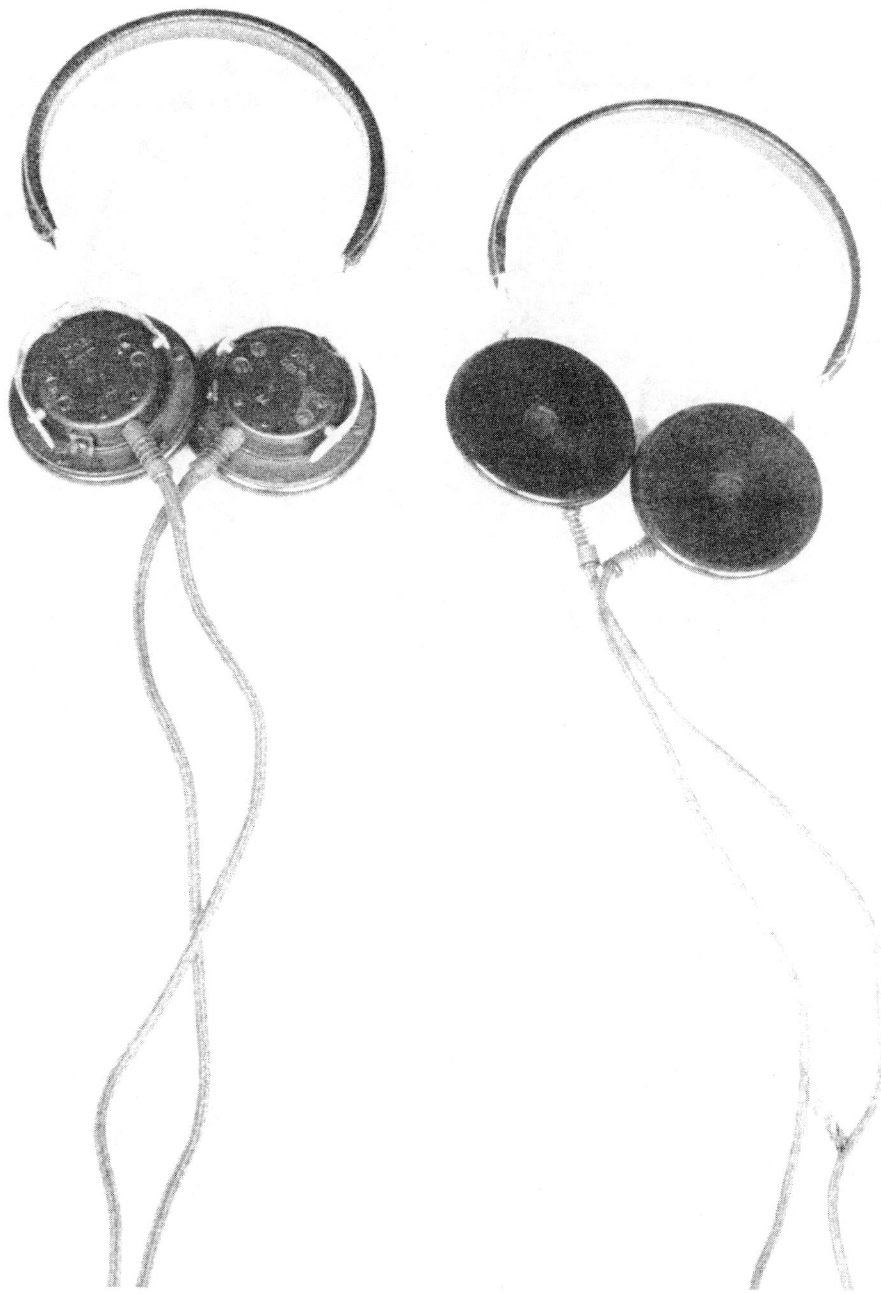

Figure 7. Standard headphones issued with the infantry radio sets, other models exist.

Figure 8. Standard armor crewman headphones. The larger rubber cups fit snuggly around the wearers ears and prevent unwanted vehicular noise.

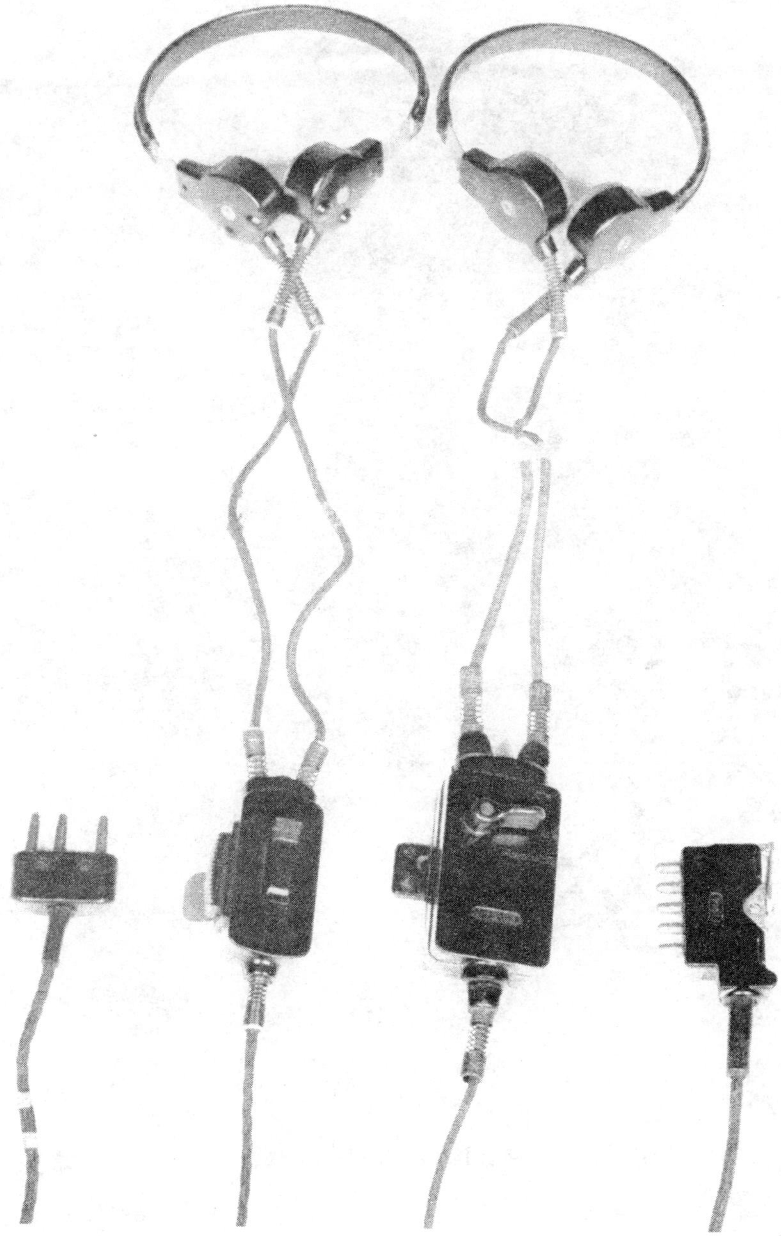

Figure 9. Two examples of the Kehlkopf (larynx) microphone. The set at the left is for radios while the one at right is for the field telephone.

Vorschriftsmäßiges Anlegen des Kehlkopfmikrofons.

Figure 10. Extract from an original manual showing use of the throat microphone.

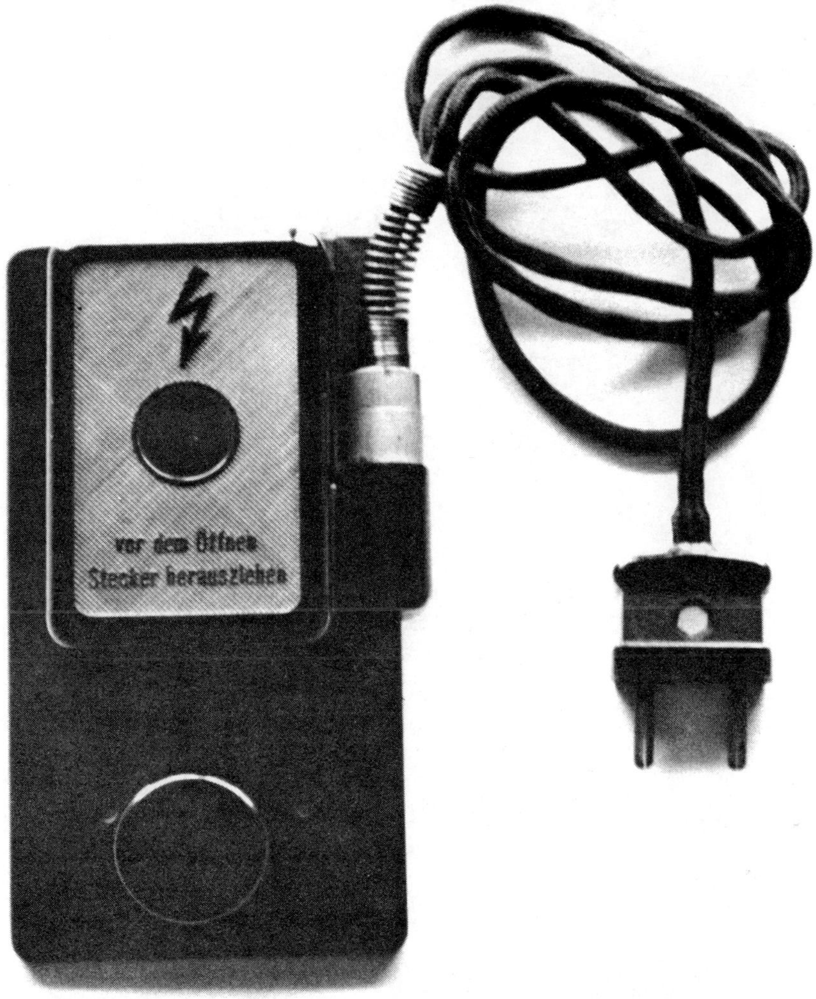

RZM Imports

Figure 11. A telegraph key with the cover closed. The warning label vor dem Offnen Stecker herausziehen translates to "unplug before opening."

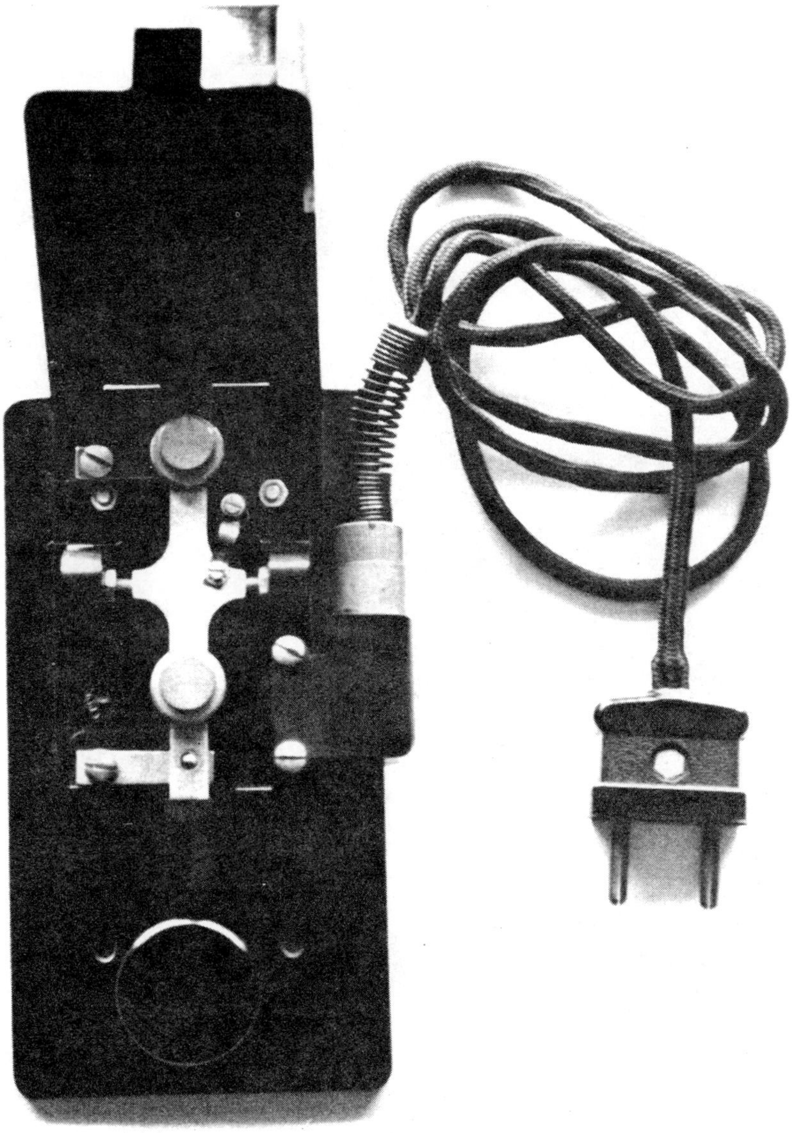

Figure 12. The same telegraph key as shown in illustration 11 with the cover open.

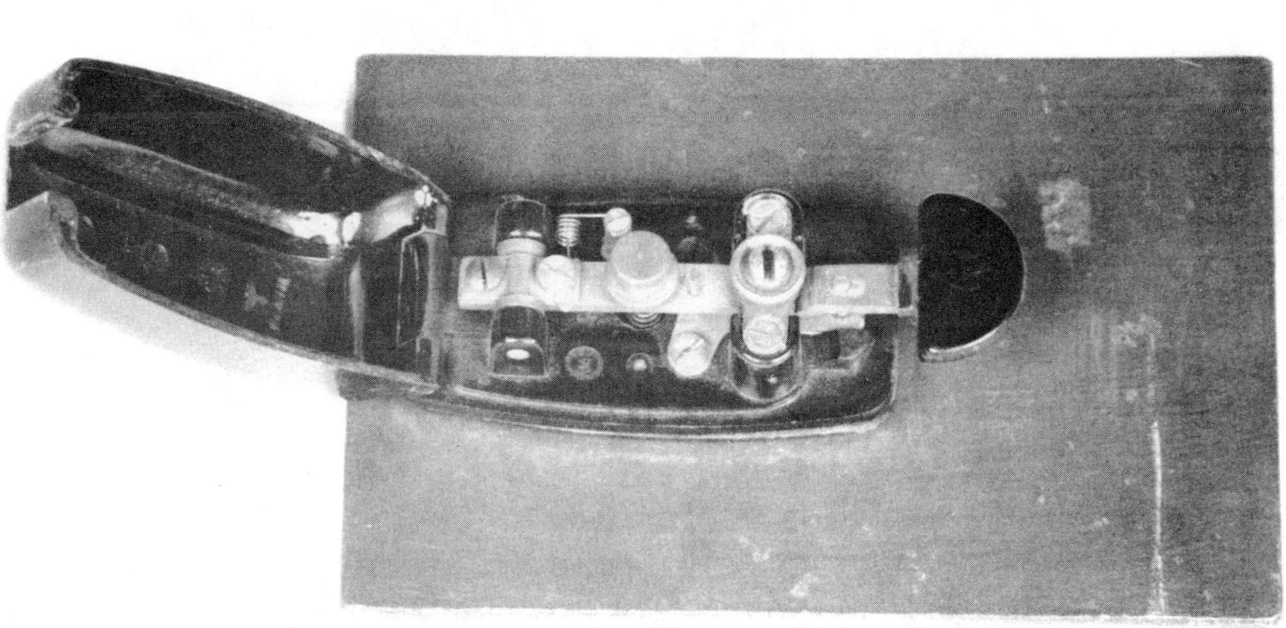

Figure 13. Another type of telegraph key known as the "mouse." This type is smaller than the example previously described. The cable is missing.

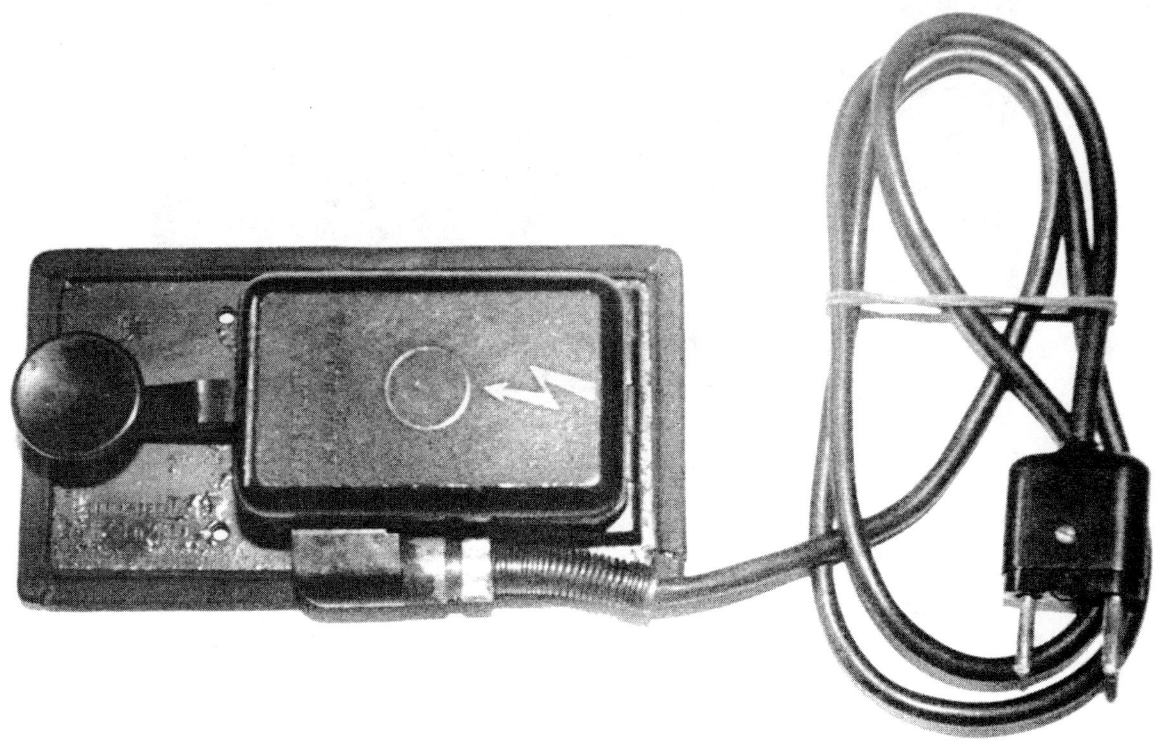

Figure 14. Similar to the previous model described in figure 11 and 12, this Morse sender has a rubber pad fitted to the base. The following manufacture mark is stamped into the metal and can be seen in the upper left corner of the base.

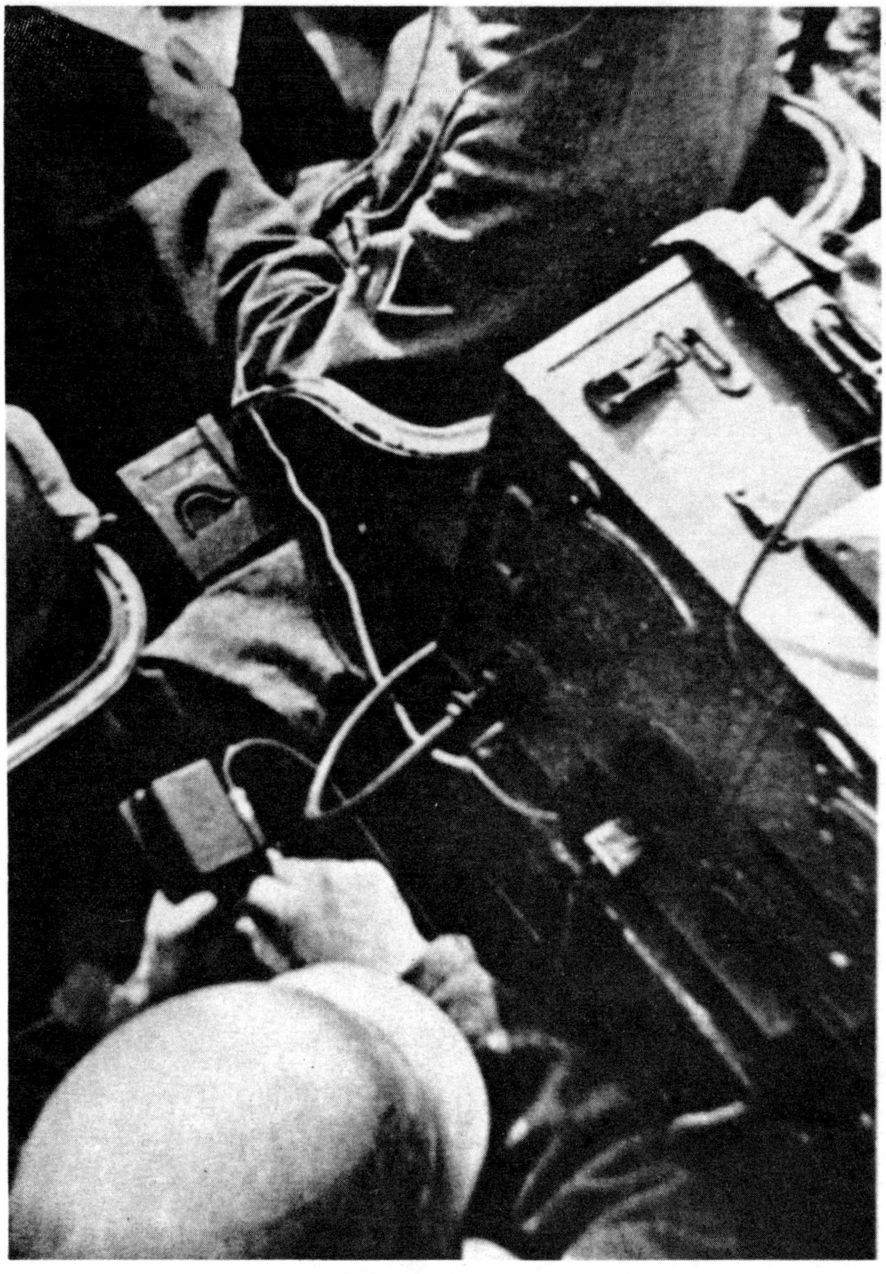

Figure 15. This radio operator is using a Morse sender (telegraph key) with a Torn. fu. bl.

Figure 16. Single headphone with fabric headband. The plug end is designed for a field telephone.

Figure 17. Hand–held cable dispenser.

Figure 18. Another view of illustration 17 showing code (hhe), year of manufacture (44), and ordnance proofmark.

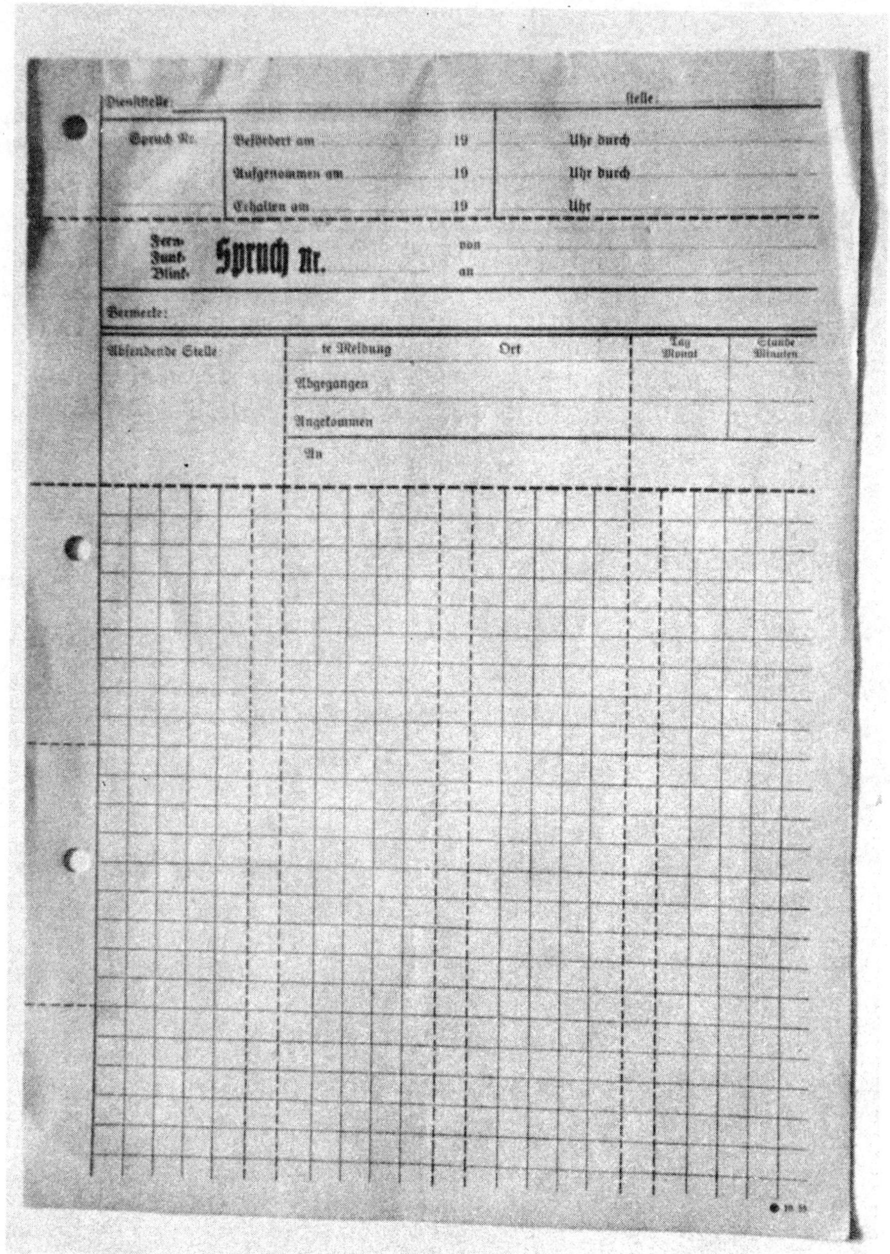

Dienststelle: stelle:

Spruch Nr.

Befördert am 19 Uhr durch

Aufgenommen am 19 Uhr durch

Erhalten am 19 Uhr

Fern-
Funk- **Spruch Nr.** von
Blink- an

Bermerke:

Absendende Stelle: te Meldung Ort Tag Monat Stunde Minuten

Abgegangen

Angekommen

An

Figure 19a. Examples of message pads. This message pad can be used for recording telephone, radio and Morse transmissions.

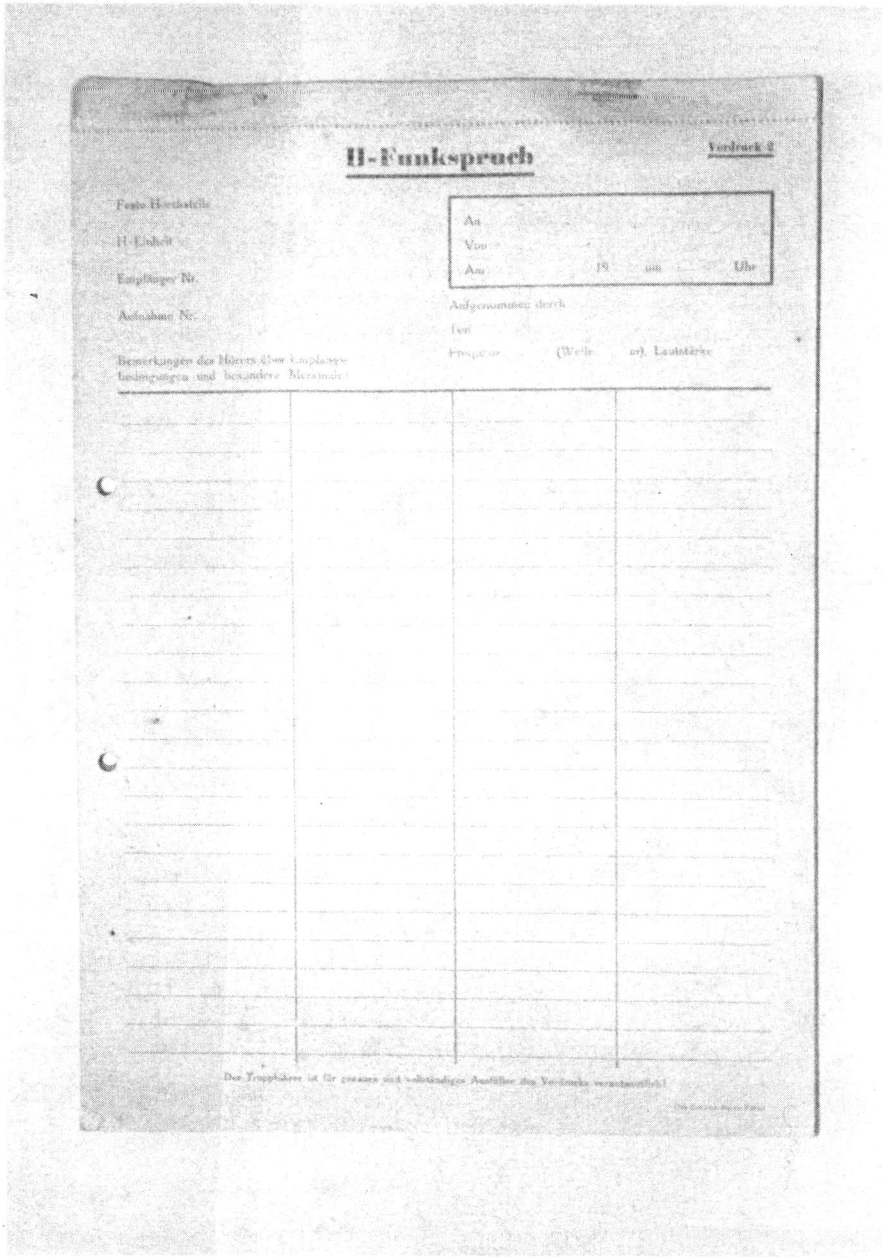

H-Funkspruch Vordruck 2

Feste Horchstelle

H-Einheit

Empfänger Nr.

Aufnahme Nr.

Bemerkungen des Hörers über Empfangsbedingungen und besondere Merkmale:

An

Von

Am 19 um Uhr

Aufgenommen durch

Frequenz (Welle m), Lautstärke

Figure 19b. This form is used for radio messages only.

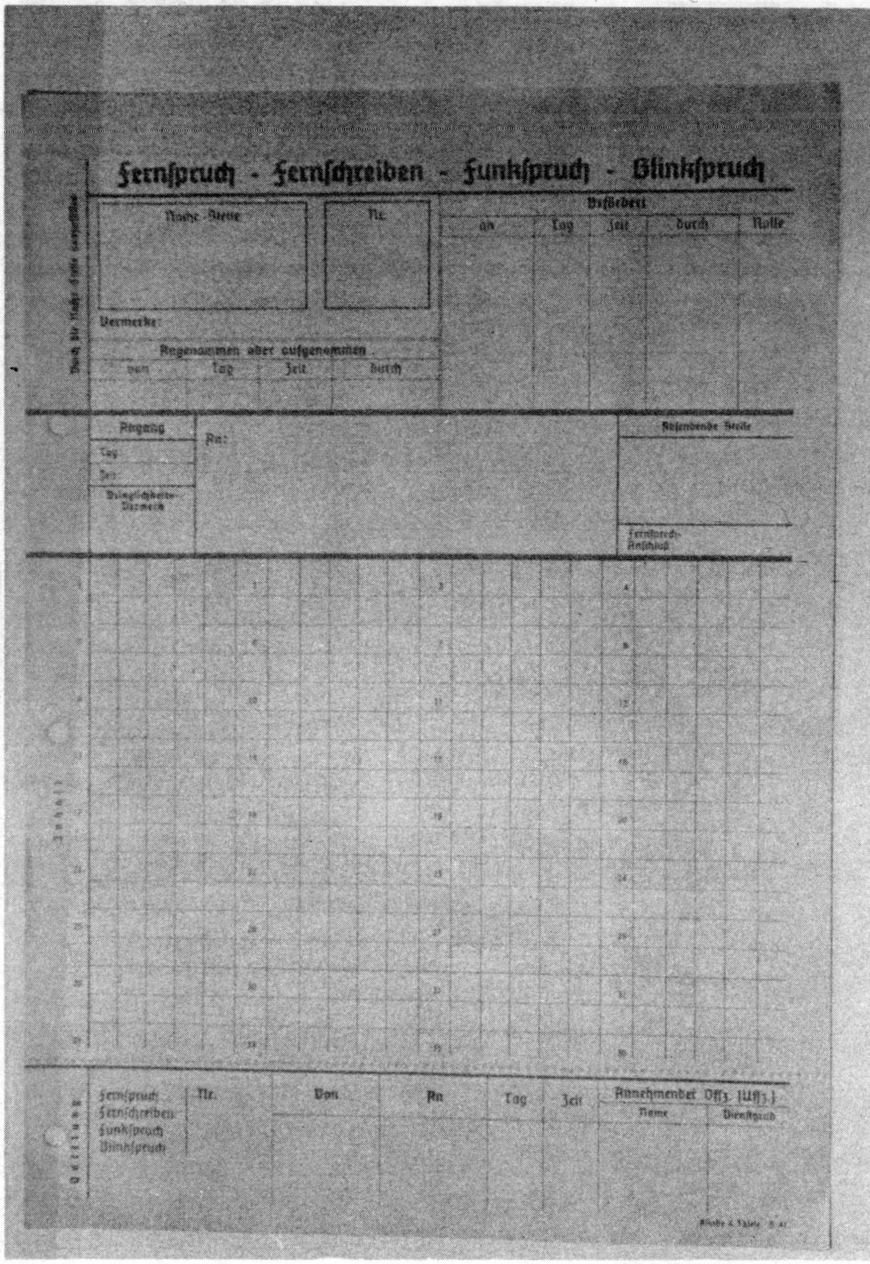

Fernspruch - Fernschreiben - Funkspruch - Blinkspruch

Nächste Stelle | Nr. | Befördert: an, Tag, Zeit, durch, Rolle

Vermerke:

Angenommen oder aufgenommen: von, Tag, Zeit, durch

Eingang: Tag, Zeit

Dringlichkeits-Vermerk

An:

Absendende Stelle

Fernsprech-Anschluß

Inhalt

Quittung: Fernspruch, Fernschreiben, Funkspruch, Blinkspruch | Nr. | Von | An | Tag | Zeit | Annehmender Offz. (Uffz.): Name, Dienstgrad

Figure 19c. This form is used for telephone, radio teletype and Morse transmissions.

Figure 20. Waffen SS signal unit prepares messages using previous shown forms. The operator in the center of the photo is using the Morse key as shown in illustration 12.

BIBLIOGRAPHY

Foreign Sources

Liste der Fertigungskennzeichen Für Waffen, Munition und Gerät; Karl R. Pawlas, Publizistisches Archiv für militär - und Waffenwesen Nurnberg 1977.

Funker am Feind; Hans Grosser; Wilhelm Limpert Verlag Berlin SW 68 1941.

I *H DV 95/13 Der Feldfernsprecher 33,* Ernst Siegfried Mittler und Sohn, G.m.b.H. 1934.

VI *D1037/5 Gebrauchsaweitung für der Kleinfunksprecher d*; Ober-Kommando des Hecres, Berlin 1944.

U.S. Sources

Ham Radio Magazine, August 1980.

III *TB SIG E-12, German Radio Sets Feldfu b and Feld fu c*, War Department, Washington, D.C., 15 May 1944 pp 1-10.

IV *TB SIG E-2, German Radio Sets Torn, fu d2,* War Department, Washington, D.C., 28 January 1944 pp 1-14.

V *TB SIG E-1, German Radio Sets Torn. Fu. b1 and Torn. Fu. f,* War Department, Washington D.C., 15 January 1944 pp 2-13.

VII *TB SIG E-9, German Radio Transmitter 100 W.S.*, War Department, Washington, D.C., 10 April 1944 pp 1-17.

VIII *TB SIG E-3, German Radio Sets Torn E.b,* E.b War Department, Washington, D.C., 10 February 1944 pp 1-14.

IX *TB SIG E-14, German Radio Receiver K.w.E.a.*, War Department, Washington, D.C., 5 June 1944 pp 1-13.

X *TB SIG E-5, German Radio Transmitters 10 W.S.c and 10 W.S.h and German Radio Receivers UKw.E.e. And UKW E h,* War Department, Washington, D.C., 8 March 1944 pp 1-16.